ENERGY POLICIES, POLITICS AND PRICES

NUCLEAR REGULATORY COMMISSION

OVERSIGHT OF THE NUCLEAR POWER INDUSTRY

ENERGY POLICIES, POLITICS AND PRICES

Additional books in this series can be found on Nova's website under the Series tab.

Additional E-books in this series can be found on Nova's website under the E-book tab.

ENERGY POLICIES, POLITICS AND PRICES

NUCLEAR REGULATORY COMMISSION

OVERSIGHT OF THE NUCLEAR POWER INDUSTRY

SHANA MAYFIELD
EDITOR

nova publishers
New York

For permission to use material from this book please contact us:
Telephone 631-231-7269; Fax 631-231-8175
Web Site: http://www.novapublishers.com

Library of Congress Cataloging-in-Publication Data

ISBN: 978-1-62948-818-9

Published by Nova Science Publishers, Inc. † New York

CONTENTS

PREFACE

The Nuclear Regulatory Commission (NRC) relies on its staff's professional judgment in implementing its processes for overseeing the safety of U.S. commercial nuclear power reactors. In implementing this oversight, NRC allocates specific roles and responsibilities to resident inspectors assigned to each plant, regional officials at one of four regional offices responsible for most oversight activities, headquarters officials, and the nuclear power industry. This book examines how NRC implements its processes for overseeing the safety of commercial nuclear power reactors; the extent to which NRC consistently identifies and resolves findings through these processes; and NRC's methods for developing lessons learned to improve its oversight and challenges, if any, NRC faces in doing so. The authors also discuss the inspector general's assessment of the most serious management and performance challenges facing the NRC.

Chapter 1 – The 2011 disaster at Japan's Fukushima Daiichi Nuclear Power Plant demonstrated that unexpected nuclear accidents with extreme consequences can occur and, thus, heightened concerns about NRC's ability to oversee the safety of U.S. commercial nuclear power reactors. NRC oversees safety through multiple processes, such as physically inspecting reactors and also responding to signs of declining performance (i.e., findings) or violations of its requirements.

GAO was asked to review NRC's oversight of the U.S. nuclear power industry. This report examines (1) how NRC implements its processes for overseeing the safety of commercial nuclear power reactors; (2) the extent to which NRC consistently identifies and resolves findings through these processes; and (3) NRC's methods for developing lessons learned to improve its oversight and challenges, if any, NRC faces in doing so. GAO reviewed

NRC policies and guidance; visited five nuclear power plants located in multiple NRC regions; analyzed NRC data on findings, violations, licensee performance, and inspection hours; interviewed NRC officials and industry representatives; and observed demonstrations of NRC's database search tools.

Chapter 2 – In accordance with the Reports Consolidation Act of 2000, I have updated what I consider to be the most serious management and performance challenges facing the U.S. Nuclear Regulatory Commission (NRC). This report conveys the updated challenge list and provides a description of each challenge area and an assessment of agency actions to address these areas.

In: Nuclear Regulatory Commission
Editor: Shana Mayfield
ISBN: 978-1-62948-818-9

Chapter 1

NUCLEAR POWER: ANALYSIS OF REGIONAL DIFFERENCES AND IMPROVED ACCESS TO INFORMATION COULD STRENGTHEN NRC OVERSIGHT*

United States Government Accountability Office

WHY GAO DID THIS STUDY

The 2011 disaster at Japan's Fukushima Daiichi Nuclear Power Plant demonstrated that unexpected nuclear accidents with extreme consequences can occur and, thus, heightened concerns about NRC's ability to oversee the safety of U.S. commercial nuclear power reactors. NRC oversees safety through multiple processes, such as physically inspecting reactors and also responding to signs of declining performance (i.e., findings) or violations of its requirements.

GAO was asked to review NRC's oversight of the U.S. nuclear power industry. This report examines (1) how NRC implements its processes for overseeing the safety of commercial nuclear power reactors; (2) the extent to which NRC consistently identifies and resolves findings through these processes; and (3) NRC's methods for developing lessons learned to improve

---

* This is an edited, reformatted and augmented version of United States Government Accountability Office, Publication No. GAO-13-743, dated September 2013.

its oversight and challenges, if any, NRC faces in doing so. GAO reviewed NRC policies and guidance; visited five nuclear power plants located in multiple NRC regions; analyzed NRC data on findings, violations, licensee performance, and inspection hours; interviewed NRC officials and industry representatives; and observed demonstrations of NRC's database search tools.

What GAO Recommends

GAO recommends, among other things, that NRC analyze the causes of differences in identifying and resolving findings across regional offices and address these differences, and that it improve its database search tools. NRC agreed with GAO's recommendations.

What GAO Found

The Nuclear Regulatory Commission (NRC) relies on its staff's professional judgment in implementing its processes for overseeing the safety of U.S. commercial nuclear power reactors. In implementing this oversight, NRC allocates specific roles and responsibilities to resident inspectors assigned to each plant, regional officials at one of four regional offices responsible for most oversight activities, headquarters officials, and the nuclear power industry. NRC also builds into its processes incentives for plant managers to identify concerns about reactor safety, report those concerns to NRC, and take prompt actions to correct them. NRC's processes for identifying and assessing findings and violations are based on prescribed agency procedures and include several points where NRC staff must exercise their professional judgment, such as determining whether issues of concern identified during physical inspections constitute findings or violations and the risk significance of any findings or the severity of any violations, among other things.

NRC is aware of differences across regional offices in identifying and resolving findings that result from physical inspections. GAO's analysis of NRC's data indicated that the number of escalated findings had fewer differences across regions than nonescalated findings, which are lower-risk findings and less severe violations. According to NRC officials, several factors, such as the hours spent on inspections, may explain the differences in nonescalated findings. However, GAO found that the regional office with the fewest reactors and the fewest inspection hours had the most nonescalated findings. NRC officials and industry representatives have raised concerns that the differences may also be due to differences in how NRC staff identify and resolve findings. NRC has taken some steps to examine the consistency of its oversight. For example, in 2009, the four regional offices implemented an initiative to explore how the regional offices identify and assess inspection findings. However, NRC has not conducted a comprehensive analysis of the causes of the differences in the number of nonescalated findings across regions. Under federal standards for internal control, managers are to compare actual performance with planned or expected results throughout the organization and analyze significant differences. Without such an analysis, NRC does not know whether its regional offices are applying regulations and guidance consistently.

NRC has both formal and informal methods for developing lessons learned to improve its oversight. Formal methods include agencywide programs, annual and biennial assessments, and special initiatives. Informal methods include reaching out to peers and technical experts across the agency and accessing various agency databases. Although NRC guidance directs inspectors to use information in agency databases on past experiences to plan and conduct inspection activities, inspectors face challenges accessing this information, which may limit their ability to use it. For example, several NRC inspectors reported contacting other inspectors informally because NRC's database search tools contain limited instructions and do not ensure thorough results. Without better search tools, inspectors may overly rely on information available through informal channels.

ABBREVIATIONS

ADAMS	Agencywide Documents Access and Management System
ALARA	as low as reasonably achievable
IG	Office of the Inspector General
INPO	Institute of Nuclear Power Operations
NFPA	National Fire Protection Association
NRC	Nuclear Regulatory Commission
ROE	Reactor Operating Experience
ROP	Reactor Oversight Process
SERP	Significance and Enforcement Review Panel

September 27, 2013

The Honorable Barbara Boxer
Chairman
Committee on Environment and Public Works
United States Senate

The Honorable Sheldon Whitehouse
Chairman
Subcommittee on Oversight
Committee on Environment and Public Works
United States Senate

The Honorable Robert Menendez
United States Senate

The Honorable Bernard Sanders
United States Senate

In March 2011, an earthquake off the Pacific coast of Japan triggered a tsunami that caused significant accidents at three reactors in the Fukushima Daiichi Nuclear Power Plant complex, demonstrating that low-probability, high-consequence nuclear accidents can and do occur.[1] According to the Japanese parliament's formal investigation, the damage to the Fukushima reactors could have been avoided and resulted, in part, from a failure of regulatory oversight in Japan. This event heightened concerns about the safety of commercial nuclear power plants worldwide and the ability of the Nuclear Regulatory Commission (NRC)—the independent federal agency responsible for licensing and regulating U.S. civilian nuclear activities—to oversee the safe operation of commercial nuclear power reactors in the United States.[2] In response to these concerns, NRC convened a Fukushima Near-Term Task Force, which in a July 2011 report concluded that a similar accident is unlikely to occur in the United States and identified many opportunities for improving NRC's oversight.[3]

NRC's mission is, in part, to ensure adequate protection of public health and safety from accidents involving commercial nuclear power reactors. In fiscal year 2012, the licensing and regulating of commercial reactors comprised over 75 percent—or approximately $800 million—of NRC's budget, of which the vast majority (87 percent) is collected in the form of fees from industry. One of NRC's stated goals is to ensure that its oversight of reactor safety is objective, predictable, and understandable, and according to NRC's strategic plan, a key strategy in achieving its mission is to conduct this oversight openly to inform the public and maintain confidence in reactor safety.

However, NRC's Office of the Inspector General (IG) reported in 2010, 2011, and 2012, that one of the most serious management and performance challenges facing NRC is the ability to effectively carry out its oversight responsibilities while responding to emerging demands associated with licensing and regulating new reactors.[4]

NRC's oversight will soon likely take on even greater importance as many commercial reactors in the United States are reaching or have reached the end of their initial 40-year operating period.[5] Nuclear power's contribution to the nation's supply of electricity—about 20 percent in 2012—will depend, in large part, on the continued operation of the nation's 100 operating commercial reactors. Most of these reactors have had their licenses renewed to operate for another 20 years.[6]

However, as we recently reported,[7] NRC's license renewal process primarily considers issues related to the negative effects of aging on reactors and their components. Other ongoing safety matters are not explicitly covered in the license renewal process because NRC considers its ongoing oversight processes as adequate to ensure that reactors are operated safely and will continue to be in the future.

We and others have reported on NRC's oversight of reactor safety. In September 2006, we reported that NRC had taken several steps to improve its oversight of nuclear reactors but had been slow to act on needed improvements, in particular, in improving the agency's ability to identify and address early indications of declining safety performance.[8] In 2008, NRC's IG found that NRC's regional offices were not consistently implementing the agency's enforcement program.[9]

Additionally, the National Academy of Sciences is conducting a study and preparing a report to NRC and Congress on lessons learned from the Fukushima nuclear accident for improving the safety and security of U.S. nuclear power plants. According to the National Academy of Sciences, this report is expected to be issued publicly by the summer of 2014.

In this context, you asked us to review NRC's oversight of the U.S. commercial nuclear power industry. This report: (1) describes how NRC implements its processes for overseeing the safety of commercial nuclear power reactors; (2) evaluates the extent to which NRC consistently identifies and resolves findings through these processes; and (3) describes NRC's methods for developing lessons learned to improve its oversight and challenges, if any, NRC faces in doing so.

To address these objectives, we reviewed relevant federal laws, regulations, and agency guidance covering NRC's processes for overseeing reactor safety.

In addition, we reviewed previous GAO and NRC IG reports that identified safety-related challenges or made recommendations to improve NRC oversight. We also reviewed NRC's efforts to update its oversight processes. We interviewed NRC headquarters officials, regional officials, and inspectors. We also visited a nonprobability sample of five selected nuclear power plants and conducted interviews with the NRC inspectors and plant managers and operators at the plants.[10] To select the five plants we visited—Calvert Cliffs Nuclear Power Plant in Maryland, Byron Station in Illinois, Palisades Nuclear Plant in Michigan, Cooper Nuclear Station in Nebraska, and Wolf Creek Generating Station in Kansas—we used multiple criteria, including whether a plant's performance (as determined by NRC) has changed (i.e., by improving or declining) in the past 2 years, and whether, collectively, plants were located in multiple NRC regions, operated by different companies that hold licenses to operate reactors (i.e., licensees), and included different reactor types. We also analyzed NRC performance data for all 65 plant sites from 2000 through 2012.

To assess the reliability of NRC's data, we reviewed existing information about the data and the system that produced them and interviewed agency officials knowledgeable about the data. We found that these data were sufficiently reliable for the purposes of our report. Our review focuses on NRC's oversight of the safe operation of reactors; physical security of plants is outside the scope of this review. A detailed description of our objectives, scope, and methodology is presented in appendix I.

We conducted this performance audit from May 2012 to September 2013 in accordance with generally accepted government auditing standards. Those standards require that we plan and perform the audit to obtain sufficient, appropriate evidence to provide a reasonable basis for our findings and conclusions based on our audit objectives. We believe that the evidence obtained provides a reasonable basis for our findings and conclusions based on our audit objectives.

BACKGROUND

This background section describes NRC's role and its processes for overseeing reactor safety, specifically the Reactor Oversight Process and the Traditional Enforcement Process.

Role of NRC

NRC is an independent agency of about 4,000 employees established by the Energy Reorganization Act of 1974 to license and regulate U.S. civilian—that is, commercial, industrial, academic, and medical—use of nuclear materials. NRC is headed by a five-member Commission. The President appoints these Commission members, who are confirmed by the Senate, and designates one of them to serve as the Chair and official spokesperson. NRC seeks to carry out its mission of ensuring safe operation of the 100 commercial nuclear power reactors currently operating in the United States from its headquarters office in Rockville, Maryland, its four regional offices—in King of Prussia, Pennsylvania (Region I); Atlanta, Georgia (Region II); Lisle, Illinois (Region III); and Arlington, Texas (Region IV)—and resident (on-site) inspectors' offices located at 62 nuclear power plants in 31 states.[11] (See fig. 1.) Appendix II provides a complete list of the commercial nuclear power reactors operating in the United States and additional information about each one. Under the Atomic Energy Act, as amended, NRC is responsible for issuing licenses to commercial nuclear reactors and conducting oversight of activities under such licenses to protect the health and safety of the public, among other things. NRC is authorized to conduct inspections and investigations, to enforce its requirements, such as by issuing orders and imposing civil (monetary) penalties, and to revoke licenses. To oversee the activities of licensed commercial reactors, NRC has established regulations, as well as guidance, including regulatory guides, standard review plans, and NRC's *Inspection Manual*. These regulations and guidance, along with plant-specific licenses and technical specifications, form the basis by which NRC provides continuous oversight of reactor operations. NRC's oversight of commercial nuclear reactors includes monitoring their performance in three strategic areas: (1) reactor safety, that is, avoiding accidents and reducing the consequences of accidents if they occur; (2) radiation safety, that is, ensuring safety from radiation for both plant workers and the public during routine operations; and (3) safeguards, that is, protecting of the plant against sabotage or other security threats. These areas comprise seven cornerstones of NRC's oversight (see table 1). According to NRC, reactors' satisfactory performance in these cornerstones provides reasonable assurance of safe operation. Our review focuses on NRC's oversight of the safe operation of reactors; therefore, the physical security of plants (Physical Protection cornerstone) is outside the scope of this review.

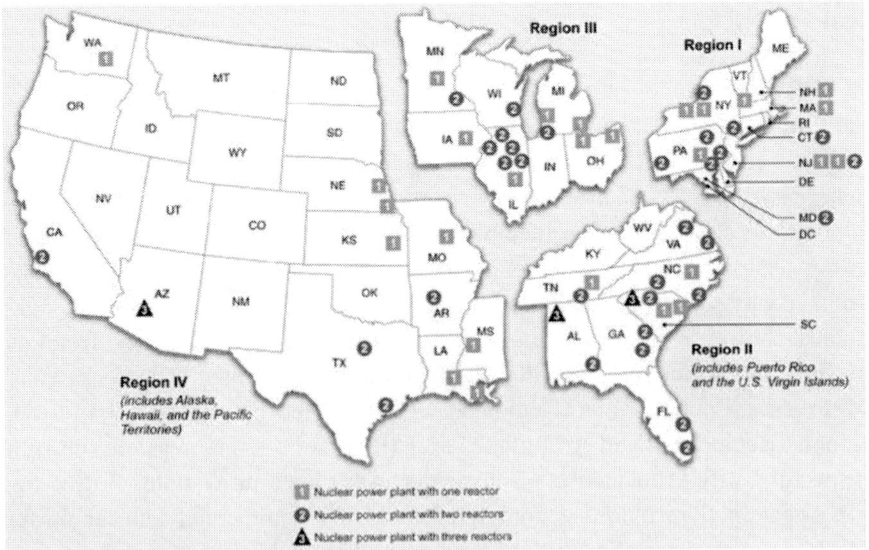

Source: GAO analysis of NRC information; Map Resources (map).

Figure 1. NRC Regions and Operating Nuclear Power Plants in the United States.

NRC also looks for systemic problems by monitoring reactors' performance in three cross-cutting areas that extend across the cornerstones:

- *Problem identification and resolution.* The licensee (1) ensures that issues potentially affecting nuclear safety are promptly identified, fully evaluated, and addressed in a timely manner; (2) uses operating experience information to support plant safety; and (3) conducts performance assessments to identify areas for improvement.
- *Human performance.* The licensee (1) makes decisions demonstrating that nuclear safety is an overriding priority; (2) ensures that staff, equipment, procedures, and other resources are available and adequate to assure nuclear safety; (3) plans and coordinates work activities

consistent with nuclear safety; and (4) uses work practices that support human performance.

- *Safety-conscious work environment.* The licensee maintains an environment in which employees feel free to raise safety concerns, both to their management and to NRC, without fear of retaliation and where such concerns are promptly reviewed, properly prioritized, and appropriately resolved.

Table 1. Strategic Areas and Cornerstones of NRC's Oversight

Strategic area	Cornerstone	Description
Reactor safety	Initiating Events	This cornerstone focuses on operations and events at a nuclear plant that could lead to a possible accident, if plant safety systems did not intervene. These events could include equipment failures leading to a plant shutdown, shutdowns with unexpected complications, or large changes in the plant's power output.
	Mitigating Systems	This cornerstone measures the function of safety systems designed to prevent an accident or reduce the consequences of a possible accident. The equipment is checked by periodic testing and through actual performance.
	Barrier Integrity	This cornerstone focuses on the physical barriers between the highly radioactive materials in fuel within the reactor and the public and the environment outside the plant. These barriers are the sealed rods containing the fuel pellets, the heavy steel reactor vessel and associated piping, and the reinforced concrete containment building surrounding the reactor. The integrity of the fuel rods, the vessel, and the piping is continuously checked forleakage, while the ability of the containment to prevent leakage is measured on a regular basis.
	Emergency Preparedness	This cornerstone measures the effectiveness of plant staff in carrying out emergency plans for responding to a possible accident. Such emergency plans are tested every two years during emergency exercises involving the plant staff and local, state, and in some cases, federal agencies.

Table 1. (Continued)

Strategic area	Cornerstone	Description
Radiation safety	Occupational Radiation Safety	NRC regulations set a limit on radiation doses received by plant workers, and this cornerstone monitors the effectiveness of the plant's program to control and minimize those doses.
	Public Radiation Safety	This cornerstone measures the procedures and systems designed to minimize radioactive releases from a nuclear plant during normal operations and to keep those releases within federal limits.
Safeguards	Physical Protection	Nuclear plants are required to have well-trained security personnel and a variety of protective systems to guard vital plant equipment, as well as programs to assure that employees are constantly fit for duty through drug and alcohol testing. This cornerstone measures the effectiveness of security and fitness-for-duty programs.

Source: GAO analysis of NRC information.

The Reactor Oversight Process

The Reactor Oversight Process (ROP) is the central component of NRC's oversight efforts. Under the ROP, NRC undertakes four key actions: (1) reviewing key indicators of licensees' performance; (2) physically inspecting reactors; (3) assessing whether any findings resulting from inspections are associated with any of the cross-cutting areas; and (4) determining what actions, if any, to take in response to signs of declining performance and any related violations of NRC requirements.

Reviews of Performance Indicators

NRC technical specialists located at its regional offices review licensee performance using quantitative measures or indicators of that performance. According to NRC documents, performance indicators are designed to illustrate a long-term view of a licensee's performance—over multiple quarters. Currently, NRC uses 17 performance indicators it developed collaboratively with the industry to measure licensee performance. Each performance indicator is designed to help gauge a licensee's performance within the different cornerstones outlined above. For example, a performance indicator under the "Initiating Events" cornerstone measures the number of unplanned reactor shutdowns during the previous four quarters. Appendix III provides a list of the safety-related (nonsecurity) performance indicators and related cornerstones.

Each licensee voluntarily submits data quarterly for each performance indicator,[12] and NRC measures licensee performance from these data against established numerical thresholds; assigning a color—green, white, yellow, or red—to reflect increasing risk to reactor safety based on these performance data. Specifically, a green indicator reflects performance within an expected range, but white, yellow, and red represent decreasing levels of a licensee's performance (see table 2).

Table 2. NRC Categories of Performance Indicators

Performance indicator category	Description
Green	Indicates performance within an expected performance level where the associated cornerstone objectives are met.
White	Represents performance outside an expected range of nominal utility performance, but related cornerstone objectives are still being met.
Yellow	Indicates related cornerstone objectives are being met, but with a moderate reduction in the safety margin.
Red	Signals a significant reduction in safety margin in the area measured by the performance indicator.

Source: NRC.

Physical Inspections

Physical inspections are one of the main tools NRC uses to oversee licensee performance. During these inspections, NRC inspectors are to verify the accuracy of the quarterly data for each performance indicator supplied by the licensee and to assess aspects of licensee performance that are not directly measured by the performance indicator data—for example, confirming that fire protection equipment is in place and that certain maintenance operations have occurred. These inspections cover a wide variety of major systems and technical areas of nuclear power reactors that NRC considers most critical to meeting the overall agency mission of ensuring nuclear power plant safety. NRC currently has 52 different inspection procedures, and they are divided into three broad types of inspections—baseline (46), supplemental (3), and reactive (3)— that vary by depth and objective. Appendix IV provides a list of inspection procedures.

Baseline inspections. Baseline inspections are the minimum level of inspections performed at all nuclear power reactors, regardless of performance, and are designed to detect declining safety performance in each of the cornerstones and to review licensee effectiveness at identifying and resolving safety problems.

NRC resident inspectors, who are located at each plant, and regional technical specialists, who travel to each plant from NRC regional offices, conduct 46 baseline inspection procedures both as needed and at intervals ranging from quarterly to triennially, involving physical observations of licensee activities and plant performance, reviewing and verifying licensee reports, and interviewing licensee personnel. Each of the baseline procedures specifies a range of sample activities to inspect. Within the scope and requirements of each procedure, inspectors select the type and number of activities to review on the basis of factors, such as the sample activities available; their risk significance; and the amount of time since a particular system or component was last inspected.[13] NRC assesses any issues of concern and determines whether to elevate them to a finding based on certain criteria.

To determine whether an issue of concern constitutes a finding, inspectors must first decide whether it is associated with a performance deficiency, which NRC guidance defines as "an issue that is the result of a licensee not meeting a requirement or standard where the cause was reasonably within the licensee's ability to foresee and correct, and therefore should have been prevented." NRC guidance defines a finding as an issue of concern associated with a performance deficiency that inspectors or regional officials then determine to be more than minor.[14]

NRC then assesses—through its ROP Significance Determination Process—the finding's risk significance.[15] Similar to performance indicators, NRC assigns the finding one of four colors to reflect increasing risk to reactor safety. Green findings—also called nonescalated findings—equate to very low risk significance, while white, yellow, and red findings—considered to be escalated findings—represent increasing levels of risk (see table 3). Appendix V provides select examples of NRC findings and related information that illustrate the varying levels of risk significance.

Table 3. NRC Categories of Risk Significance for Inspection Findings

Inspection finding category	Risk significance	Description
Green	Very low safety significance	Indicates that licensee performance is acceptable and cornerstone objectives are fully metwith nominal risk and deviation.
White	Low to moderate safety significance	Indicates that licensee performance is acceptable, but outside the nominal risk range. Cornerstone objectives are met with minimal reduction in safety margin.
Yellow	Substantial safety significance	Indicates a decline in licensee performance that is still acceptable with cornerstone objectives met, but with significant reduction in safety margin.
Red	High safety significance	Indicates a decline in licensee performance that is associated with an unacceptable loss of safety margin. Sufficient safety margin still exists to prevent undue risk to public health and safety.

Source: NRC.

Supplemental inspections. When NRC identifies one or more escalated findings or a performance indicator exceeding the green threshold at a reactor, regional officials conduct supplemental inspections, which expand the scope beyond baseline inspection procedures and focus on diagnosing the cause of the finding. There are three levels of supplemental inspections that increase in breadth and depth of analysis, depending on the number and type of performance problems identified. The first and lowest level of supplemental inspection assesses the licensee's corrective actions to ensure they were sufficient in both correcting the problem and addressing the root and contributing causes to prevent recurrence. The second level of supplemental inspection has an increased scope that includes independently assessing the extent of the condition for both the specific and any broader performance problems. The third and highest level of supplemental inspection is more comprehensive and includes determining whether the reactor can continue to operate and whether additional regulatory actions are necessary. The highest level of supplemental inspection is usually conducted by a multidisciplinary team of NRC inspectors and may take place over several months. Also, as a part of this inspection, NRC inspectors assess the adequacy of the licensee's overall programs for identifying, evaluating, and correcting performance issues.

Reactive inspections. NRC conducts reactive inspections of licensee performance when specific events occur that are of particular interest to NRC because of their potential safety significance or because of potential generic safety concerns important to multiple licensees. Reactive inspections determine the cause of the event and assess the licensee's response to the event. Reactive inspections have three levels that are normally determined by an event's safety, or risk significance. According to NRC officials, recent reactive inspections have focused on issues such as an automatic reactor shutdown following a loss of offsite power, and a flaw discovered in a weld on a reactor component. To conduct reactive inspections, a team of experts is formed and an inspection charter issued that describes the scope of the inspection efforts.

Assessments of Findings for Cross-cutting Areas

NRC inspectors assess all of the findings that result from physical inspections to determine if any of the three cross-cutting areas contributed to a performance problem. If regional officials determine that one of the cross-cutting areas contributed to a finding, NRC assigns what it calls a cross-cutting aspect within that area to the finding. More than one cross-cutting aspect may be assigned to a finding, but this does not occur often, according to NRC officials. Every 6 months, NRC regional officials analyze all findings issued at each plant during the prior 12- month assessment period. In general, if a nuclear plant has four findings with the same cross-cutting aspect assigned during the previous year, and NRC is concerned about the licensee's progress in addressing them, NRC determines that the nuclear plant has what NRC calls a substantive cross-cutting issue.[16] According to NRC documents, the agency identifies substantive cross-cutting issues to inform the licensee that it has a concern with its performance and to further encourage it to take appropriate actions before more significant performance issues emerge.

Response to Performance Indicators and Findings

NRC assesses its reviews of performance indicators and all findings that result from its physical inspections to determine what actions, if any, to take in response to signs of declining performance, including violations of NRC requirements.[17]

Table 4. NRC Reactor Oversight Process Action Matrix

Oversight category:	Licensee response (column 1)	Regulatory response (column 2)	Degraded cornerstone (column 3)	Multiple/repetitive degraded cornerstone (column 4)	Unacceptable performance (column 5)
NRC's assessment of licensee performance by reactor:	All green findings and performance indicators	One white finding or performance indicator, or two white findings or performance indicators in different cornerstones	Two or more white findings or performance indicators in one cornerstone, or one yellow finding or performance indicator, or any three white findings or performance indicators	Two white findings or performance indicators or one yellow finding or performance indicator in one cornerstone for five or more quarters, or multiple yellow or one red finding or performance indicator	Overall unacceptable performance representing situations—such as multiple significant violations of a plant's license—in which NRC does not have reasonable assurance that the licensee can or will conduct its activities to ensure protection of public health and safety. In general, it is expected, but not required, that entry into column 4 will precede consideration of whether a plant is in column 5.
NRC's oversight actions:	Baseline inspections only	Baseline inspections and first level of supplemental inspections	Baseline inspections and second level of supplemental inspections	Baseline inspections and third level of supplemental inspections[a]	Order to modify, suspend, or revoke licensed activities
NRC office responsible for oversight actions:	Regional office	Regional office	Regional office	Agency[b]	Agency[b]

Source: GAO analysis of NRC information.

Note: In addition to the actions listed in this table, increasingly higher levels of NRC management will meet with a licensee as it moves to the right on the action matrix.

[a] For reactors at this oversight level, at a minimum, the licensee and NRC are to document agreement on the corrective actions the licensee will take through a performance improvement plan. NRC may also take actions including making a demand to the licensee for information or issuing an order up to and including a plant shutdown.

[b] Agency response involves senior management attention from both headquarters and regional offices.

The significance of any violations associated with a finding is incorporated into the color coded risk assessment made during the Significance Determination Process. For enforcement of violations associated with nonescalated findings, NRC establishes a public record of the violations and requires the licensee to take steps to correct the violations, but it does not usually require the licensee to document its responses. For enforcement of violations associated with escalated findings, NRC issues written notices of violation (to which licensees are typically required to respond), civil penalties, or orders.[18]

Based on NRC's assessment of licensee performance under the ROP, NRC places each of the licensee's reactors into one of five performance categories on its action matrix, which corresponds to graded, or increasing, levels of oversight.

The action matrix is NRC's formal method of determining how much additional oversight—mostly in the form of supplemental inspections—is required on the basis of the number and risk significance of performance indicators and inspection findings.[19] NRC considers licensee performance to be acceptable until it reaches column 5 (see table 4).

On the basis of the results of the ROP, NRC provides licensees and the public with an overall assessment of each reactor's performance. At the end of each 6-month period, NRC issues an assessment letter to each plant to describe what level of oversight the plant's reactors will receive according to their placement on the action matrix, what actions NRC is expecting the licensee to take as a result of the performance issues identified, any specific enforcement actions NRC has taken, and any documented substantive cross-cutting issues. If a substantive crosscutting issue is identified, the letter will describe what actions NRC intends to take to monitor the issue. If the issue persists, NRC can request a number of actions from the licensee, including a written response describing the corrective actions the licensee intends to take. NRC also holds an annual public meeting at or near each plant to review its performance and address questions from members of the public and other interested stakeholders.

Additionally, NRC has other mechanisms to make available its oversight results, such as a website that provides summaries of each reactor's current overall performance.[20] While it addresses most violations through the ROP, NRC also responds to some violations identified during the ROP under what it calls its Traditional Enforcement Process.[21]

NRC's Traditional Enforcement Process

NRC uses this process to assess and determine the enforcement actions, if any, to take in response to violations that, according to NRC guidance, cannot be addressed fully through the ROP. The process typically applies to violations that result in actual safety consequences, may impede NRC's oversight of licensed activities, or involve deliberate misconduct or discrimination matters, as well as certain technical violations involving specific areas, such as operator licensing and fuel storage. The Traditional Enforcement Process, including possible civil penalties, is managed outside the ROP action matrix.[22]

The Traditional Enforcement Process focuses on a violation's consequences and results in the assignment of a significance level ranging from Severity Level IV for the least significant to Severity Level I for the most significant concerns, according to NRC policy. NRC's response to Severity Level IV violations is considered "nonescalated enforcement," while its response to Severity Levels I, II, and III is considered "escalated enforcement." In assigning a Severity Level to the violation, NRC assesses the following: (1) the actual safety consequences; (2) the potential safety consequences; (3) the potential for impacting NRC's ability to perform its regulatory function (e.g., failure to provide complete and accurate information); and (4) any willful aspects of the violation. Severity Levels I and II violations generally involve actual or high potential consequences to public health and safety. Severity Level III violations are cause for significant concern, and Severity Level IV violations are less serious but are of more than minor concern. Appendix V provides select examples of violations that illustrate varying levels of severity. NRC includes results from the Traditional Enforcement Process in the overall semiannual assessment of each plant's performance that it provides to plant licensees and the public.

NRC Relies on Staff's Professional Judgment in Implementing Its Processes for Overseeing the Safety of Commercial Nuclear Reactors

NRC relies on its staff's professional judgment in implementing its processes, which are largely prescribed in guidance, for overseeing the safety of commercial nuclear power reactors. In implementing its oversight programs, NRC allocates specific roles and responsibilities to resident inspectors, regional staff, headquarters officials, and the nuclear power industry. NRC also provides incentives for licensees to identify and report any concerns about reactor safety to NRC. NRC oversight relies on its staff's professional judgment in applying regulations and guidance, such as determining: (1) whether issues of concern identified in inspections constitute findings, (2) the risk significance of any findings, and (3) whether these findings have cross-cutting aspects, among other things.

NRC Allocates Specific Oversight Roles and Responsibilities among Regional Officials, Resident Inspectors, Headquarters Officials, and the Nuclear Power Industry

To implement its oversight, NRC allocates specific roles and responsibilities to regional officials, resident inspectors, headquarters officials, and the nuclear power industry. According to NRC officials, doing so allows decisions to be made at the appropriate level within the agency, but it also necessitates open communication between the various offices to effectively share information and ensure policies and processes are implemented consistently across the agency.

Regional Officials

Officials from NRC's four regional offices are responsible for implementing the ROP and the Traditional Enforcement Process and establish—consistent with agency guidelines—the region's procedures for assessing issues of concern and resolving most findings.[23] According to NRC documents, the inspection program is intended to provide regional administrators flexibility with the planning and application of inspection resources to deal with risk-significant issues and problems. According to NRC officials, regional branch chiefs supervise small groups of resident inspectors, facilitate information sharing among inspectors in their region, and manage specialized teams of inspectors that conduct periodic plant inspections focused on specific technical areas, such as fire safety. Further, NRC officials said that regional branch chiefs are largely responsible for deciding whether an issue of concern raised during physical inspections constitutes a finding. Agency guidance provides them with relative autonomy in determining when an issue constitutes a more-than-minor performance deficiency.[24]

Since NRC established the ROP in 2000, regional offices have been responsible for addressing all nonescalated findings and violations, which constituted about 90 percent of findings and violations, according to our analysis of NRC data. In addition, though subject to oversight by NRC headquarters, regional offices play a lead role when assessing and responding to escalated findings or violations, including requiring licensees to provide a written response describing the reasons for a performance deficiency, actual and proposed corrective actions, and the expected date of compliance. After assessing licensee performance based, in part, on the findings or violations, regional offices then conduct supplemental inspections as appropriate, according to NRC guidance.

Resident Inspectors

NRC assigns resident inspectors to each nuclear power plant to conduct baseline inspections.[25] According to NRC guidance, resident inspectors are responsible for routinely conducting plant status reviews that include observing control room activities; attending licensee meetings; and conducting walk-downs, or inspections, of various plant areas. In addition to these duties, resident inspectors are responsible for ensuring that plant representatives implement corrective actions within a "reasonable period of time" in response to nonescalated findings.[26]

Headquarters Officials

NRC headquarters officials play a key role in addressing escalated findings and violations, those assigned the highest categories of risk significance or severity levels. For example, NRC officials told us that, before findings are assessed as being escalated, NRC headquarters officials participate in meetings with regional officials and resident inspectors to ensure that the assessment is accurate and any proposed enforcement actions are justifiable. Moreover, according to NRC guidance, officials in NRC headquarters must approve all enforcement actions against licensees. Specifically, this includes the Director of NRC's Office of Enforcement, and, in certain cases, such as when responding to Severity Level I violations, the Deputy Executive Director, after consultation with the Commission. These actions can include anything from monetary penalties, to orders to take specific actions, to shutting down a plant.

Nuclear Power Industry

In addition to NRC, the commercial nuclear power industry also plays a key role to ensure that nuclear plants operate safely and comply with regulations. According to NRC policy, licensees have the primary responsibility for safely operating their plants, as well as in accordance with their licenses and NRC regulations. Each plant has many physical structures, systems, and components, and licensees have numerous activities under way 24 hours a day to ensure that plants operate safely. These plants were designed according to a "defense-in-depth" philosophy revolving around redundant, diverse, and reliable safety systems. For example, two or more key safety components are in place so that if one fails, another one is present to serve as a backup. These include safety components such as safety injection pumps, sources of power, and physical barriers to contain radiation.

Furthermore, the Institute of Nuclear Power Operations (INPO)—the industry's self-regulator—is a nonprofit organization formed in 1979 by the nuclear power industry in response to the accident at Three Mile Island.[27] INPO has a key role in ensuring that plants operate safely by seeking to promote high levels of safety and reliability in the operation of nuclear power plants. According to its website, INPO sets performance objectives, criteria, and guidelines for the nuclear power industry, conducts evaluations of nuclear power plants, and communicates lessons learned and best practices throughout the nuclear power industry. According to INPO information, all organizations with direct responsibility to operate or construct commercial nuclear power plants in the United States have maintained continuous membership in INPO since its inception, and are primarily responsible for its funding. INPO provides a variety of services to the nuclear power industry, including personnel training for key positions at nuclear power plants, periodic evaluations and peer reviews of operating plants that are focused on plant safety and reliability, and assessments of plant performance across the industry to help licensees to understand issues that arise at other plants. According to INPO information, the organization has conducted nearly 1,200 plant evaluations, inspecting each plant approximately every 2 years, since 1980. Also, according to an INPO official, INPO uses peer pressure to influence plants to improve, and in an industry presentation, its president and chief executive officer stated that its policy of providing licensees with confidential evaluation reports allows it to have more open and candid discussions with plants about problems and areas for improvement. In addition, the industry's collective insurance company for operational losses—Nuclear Electric Insurance Limited—uses INPO ratings as a factor in setting insurance premiums for insuring for costs associated with certain long-term interruptions of electricity supply and damages to plants, among other things.[28] According to INPO and NRC documents, INPO and NRC communicate on topics of mutual interest related to improving the performance of commercial nuclear power plants.

NRC also builds into its processes incentives for licensees to identify concerns about reactor safety, discuss those concerns with NRC, and identify actions to correct them. According to NRC guidance, these concerns—known as licensee-identified findings—include any findings the licensee is already addressing through the plants' respective corrective action programs. If NRC's preliminary assessment of a licensee-identified finding indicates that the finding is green, NRC does not consider it when assessing licensee performance, does not consider it to count toward substantive cross-cutting issues, and does not apply additional oversight or require additional response from the licensee. According to NRC officials, licensee-identified findings demonstrate that the licensee is taking a proactive approach and can identify and correct problems before NRC identifies them; therefore NRC gives licensees credit for identifying potential problems during their own inspection processes. Regarding licensee-identified violations, according to NRC's enforcement policy, NRC should normally give the licensee "credit" when assessing for civil penalties—that is, the violation should be issued with no civil penalty—for identifying the violation and taking prompt and comprehensive corrective actions. On the other hand, the policy states that NRC should issue a civil penalty, regardless of whether the violation is licensee-identified, if NRC officials determine that the corrective actions have not been prompt and comprehensive.

Conversely, according to its enforcement policy, NRC considers some findings and violations—whether identified by a licensee or an inspector— to be "self-revealing." For example, according to an agency official, a licensee may discover a pool of oil on the floor next to a diesel generator, indicating that something is wrong with the equipment. NRC may consider such a finding to be self-revealing, even if the licensee identified it. Under NRC's enforcement policy, it may give the licensee credit for such findings under certain circumstances, such as licensee effort in discovering the root cause of the finding.

NRC Oversight Relies on Its Staff's Professional Judgment in Applying Regulations and Guidance

NRC's oversight processes are based in regulation, and its specific processes for identifying and assessing findings and violations are largely prescribed in guidance, but rely on several key points where NRC staff must exercise their professional judgment, such as determining (1) whether an issue of concern constitutes a finding or violation, (2) a finding's risk significance or a violation's severity, and (3) whether findings or violations have cross-cutting aspects, among other things.

Determining What Constitutes a Finding

According to NRC documents and officials, NRC inspectors use their professional judgment and a combination of education, experience, and prescribed agency procedures to make initial assessments as to whether an issue of concern identified during an inspection constitutes a finding. NRC officials told us that performance deficiency was defined broadly to allow officials to use their professional judgment when assessing issues of concern. If NRC inspectors determine that there is a performance deficiency, they and regional staff then determine whether the performance deficiency is more than minor. In doing so, these officials have NRC-established questions and guidance and use their professional judgment in applying them. If inspectors and regional staff determine that a performance deficiency is not more than minor, then it is not considered a finding and the matter is not pursued any further under the ROP. To help in making these determinations, inspectors can request additional information from licensees and consult NRC inspectors assigned to other nuclear plants, according to NRC officials. If resident inspectors conclude they cannot adequately assess an issue of concern, they can seek assistance from their regional office or NRC headquarters—typically from technical specialists or branch chiefs who supervise resident inspectors at two to three plants—to help make a determination. As we previously reported,[29] inspectors also select the type and number of activities to review during inspections, in part, on the basis of an inspector's professional judgment, which is based on information such as reviews of the licensee's corrective action program, allegations, or interviews with plant employees.

Determining a Finding's Risk Significance and the Severity of a Violation

Professional judgment also plays a key role in regional officials' determining the risk significance of a finding and the severity of a violation. If NRC determines that a performance deficiency is more than minor, it is considered a finding, and NRC regional officials assess its significance using the Significance Determination Process, which also requires the staff to exercise their professional judgment, according to NRC officials. This process provides an initial screening to identify those nonescalated findings—that is, those findings that do not result in a significant increase in plant risk and thus need not be analyzed further. Remaining findings are subjected to additional reviews to determine their risk significance, using the next phase of the Significance Determination Process—the Significance and Enforcement Review Panel (SERP).[30] According to NRC guidance, this more detailed assessment may involve NRC risk experts from the appropriate regional office or NRC headquarters and further review by the licensee's plant staff. The final outcome of the ROP review—evaluating whether the finding is green, white, yellow, or red—is used to determine NRC actions that may be needed. For any escalated violations identified during the ROP that are to go through the Traditional Enforcement Process, the severity level of violations is determined using a combination of NRC's enforcement policy, professional judgment, and information obtained from the licensee during a conference with them, if necessary. NRC officials initially determine enforcement actions during an enforcement panel, and then make a final determination during a meeting, or caucus, both of which loosely correspond to the SERP.

Regarding violations, NRC's enforcement policy states that, because the regulation of nuclear activities does not lend itself to a mechanistic treatment in many cases, judgment and discretion must be exercised in determining the severity levels of violations and the appropriate enforcement sanctions. According to this policy, the use of judgment and discretion includes the decision to issue a notice of violation or to propose or impose a civil penalty and the amount of the penalty, after considering the general principles of NRC's enforcement policy and the significance of the violation, as well as the surrounding circumstances. According to its enforcement policy, NRC may choose to exercise discretion and either escalate or mitigate enforcement sanctions or otherwise refrain from taking enforcement action within the Commission's statutory authority. According to this policy, NRC may exercise judgment and discretion to determine the severity levels of violations and the appropriate enforcement sanctions to be taken. For example, NRC may refrain from issuing a notice of violation or a proposed civil penalty for a Severity Level II, III, or IV violation that is identified after NRC has taken enforcement action, if the violation is identified by the licensee as part of the corrective action for the previous enforcement action and the violation has the same or similar root cause as the violation for which enforcement action was previously taken, among other things.

Determining Whether Findings Have Cross-cutting Aspects

In addition, according to NRC guidance, inspectors should also use their professional judgment when assessing whether a finding is associated with any cross-cutting aspects. For example, as we reported,[31] in analyzing the failure of a valve to operate properly at one plant, NRC inspectors determined that plant employees had not followed the correct maintenance procedures; thus, the finding was associated with the human performance cross-cutting area.

Both opening and closing a substantive cross-cutting issue requires the use of professional judgment. According to NRC guidance, before opening a substantive cross-cutting issue, regional officials must determine, among other things, whether the licensee is already taking actions to address the issue and, if so, whether these actions are timely, are commensurate with its significance, and will be effective in substantially mitigating it. Once a substantive cross-cutting issue has been opened, the regional office that opened it is responsible for establishing the criteria for closing it. These criteria may include having fewer findings with the same cross-cutting area in a subsequent assessment period or the regional staff's increased confidence in the licensee's ability to address the substantive cross-cutting issue. The decision to close a substantive cross-cutting issue is made by regional staff, including the senior resident inspector, branch chief, and other regional management, but NRC headquarters officials told us they also participate.

NRC can require licensees to take several actions to close substantive cross-cutting issues. According to NRC guidance, after first issuing a letter identifying a substantive cross-cutting issue to a licensee, NRC staff may follow up on these issues in multiple ways, including semiannual evaluations or additional inspections. In the second consecutive assessment letter identifying the same substantive cross-cutting issue, the regional office may consider requesting the licensee provide a response at a public meeting or provide a written response or hold a separate meeting. In the third consecutive letter, according to NRC guidance, the regional office would typically request that the licensee perform an independent assessment of safety culture, although the regional office could decide that a safety culture assessment is not necessary if the licensee has made reasonable progress addressing the issue. If the substantive cross-cutting issue continues beyond the third letter, the regional office may consider additional actions not prescribed by the action matrix developed in consultation with the Director of NRC's Office of Nuclear Reactor Regulation and its Executive Director of Operations.[32]

Figure 2 outlines NRC's processes for identifying and assessing findings and violations and key points where professional judgment plays a role.

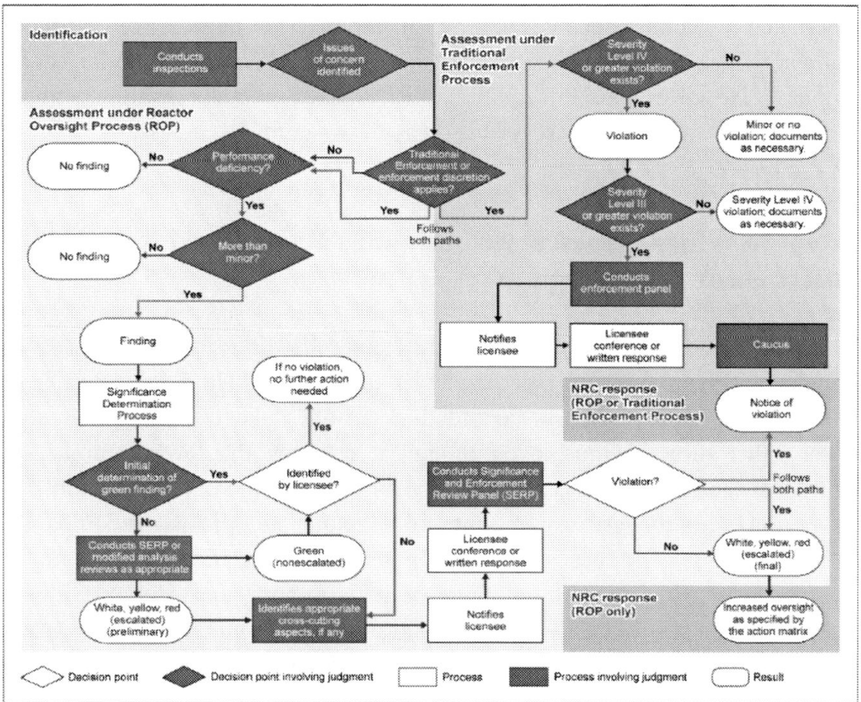

Source: GAO analysis of NRC information.

Figure 2. NRC's Processes for Identifying and Assessing Findings and Violations at Nuclear Power Reactors.

Differences Exist across NRC Regions in Identifying and Resolving Findings, and NRC Has Taken Some Steps to Address Them

NRC is aware of differences across regions in identifying and resolving the findings that result from its oversight processes and has taken some steps to address them. In particular, the extent to which nonescalated findings, which equate to very low risk significance, have been identified differs across regions.[33] Several factors may explain these differences, but NRC officials and industry representatives have raised concerns that they may be due, in part, to differences in how NRC staff identify and resolve findings. The four NRC regions also differed in the number of cross-cutting aspects they assigned to findings. NRC has taken some steps to address these differences but has not undertaken a comprehensive review of them. The number of escalated findings, which equate to greater risk significance, were more similar across regions.[34] However, the information NRC makes publicly available about findings and how they are resolved is difficult to search and assess.

Extent to Which Nonescalated Findings Have Been Identified Differs across NRC Regions

Some regions identify more nonescalated findings than others, even though they oversee fewer reactors. For example, as table 5 indicates, from 2000 through 2012, Region IV recorded more nonescalated findings than each of the other three regions, even though it oversaw the fewest number of reactors at the fewest number of plants. During the same period, Region II—the region with the greatest number of reactors and plants—recorded the fewest number of nonescalated findings. More specifically, Region IV recorded over 70 percent more nonescalated findings than Region II even though Region IV oversaw over one-third fewer reactors; resulting in an average number of nonescalated findings per reactor for Region IV to be over 2.6 times that of Region II.

Table 5. Nonescalated Findings per NRC Region, Calendar Years 2000-2012

Nonescalated findings, plants, and reactors	Region I (Northeast)	Region II (Southeast)	Region III (Midwest)	Region IV (West)
Number of nonescalated findings	2,518	1,885	3,148	3,225
Number of plants	17	18	16	14
Number of reactors	26	33	24	21
Average number of nonescalated findings per reactor	96.8	57.1	131.2	153.6

Source: GAO analysis of NRC data.

Note: This table includes data on the 104 commercial nuclear reactors operating in the United States through the end of calendar year 2012. In February 2013, the owner of the Crystal River Nuclear Plant in Florida permanently shut down that plant's reactor; in May 2013, the owner of the Kewaunee Power Station in Wisconsin permanently shut down that plant's reactor; and in June 2013, the owner of the San Onofre Nuclear Generating Station in California permanently shut down that plant's two reactors. These actions reduced the number of operating commercial nuclear reactors in the United States from 104 to 100.

As figure 3 shows, from 2000 through 2012, the number of nonescalated findings identified differed across NRC's four regions, but Region IV recorded more nonescalated findings than any of the other three regions for 6 consecutive years—2007 through 2012. Appendix VI lists the number of nonescalated findings identified at each plant from 2000 through 2012.

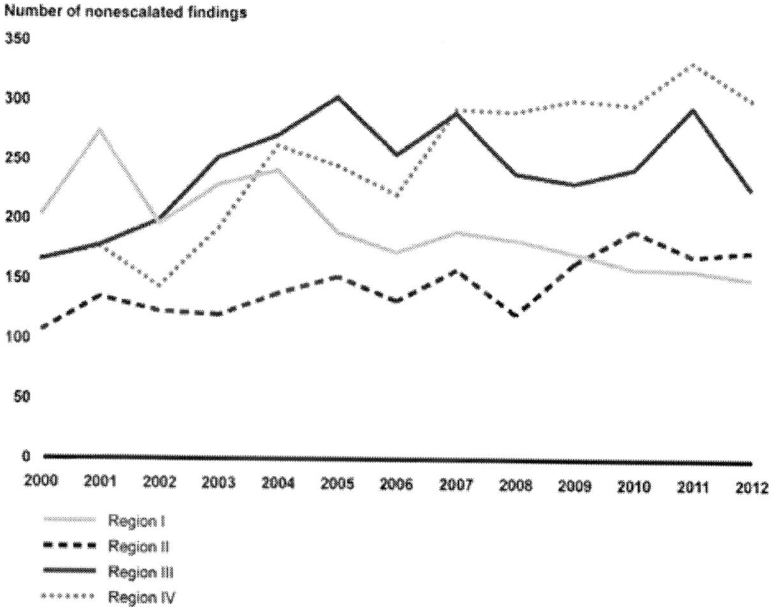

Source: GAO analysis of NRC data.

Note: Region I oversees the Northeast; Region II oversees the Southeast; Region III oversees the Midwest; and Region IV oversees the West.

Figure 3. Nonescalated Findings by NRC Region and by Year, Calendar Years 2000-2012.

Various Factors May Contribute to Differences in the Number of Findings Identified

NRC officials stated that various factors may contribute to differences in the number of nonescalated findings identified across regions, including differences in the (1) number of hours spent inspecting reactors; (2) amount of time reactors are under increased oversight as a result of performance deficiencies; (3) age of reactors; and (4) benefits from the combined resources of multiple plants owned or managed by one licensee.

Differences in inspection hours. NRC officials stated that differences in the number of hours regions spent inspecting reactors may have contributed to the differences in the number of nonescalated findings across regions, because more inspection hours would lead to more findings. However, according to 2000 through 2012 NRC data, while the total number of inspection hours has varied across the regions, the number of inspection hours has been fairly consistent across regions when accounting for the number of reactors in each region. When comparing the number of nonescalated findings in each region to the number of inspection hours and accounting for the regional differences in the number of reactors, we found that the number of nonescalated findings per inspection hour varies widely. Specifically, Region II identified the least number of nonescalated findings per inspection hour per reactor, and Region IV identified the most, with more than 3.5 times as many nonescalated findings per 1,000 inspection hours per reactor as Region II (see table 6).

Table 6. Nonescalated Findings and Inspection Hours per NRC Region, Calendar Years 2000-2012

Nonescalated findings, inspection hours, and reactors	Region I (Northeast)	Region II (Southeast)	Region III (Midwest)	Region IV (West)
Number of nonescalated findings	2,518	1,885	3,148	3,225
Number of inspection hours (thousands)	562.8	532.2	505.3	396.1
Number of reactors	26	33	24	21
Number of inspection hours (thousands) per reactor	21.6	16.1	21.1	18.9
Number of nonescalated findings per 1,000 inspection hours	4.5	3.5	6.2	8.1
Average number of nonescalated findings per 1,000 inspection hours per reactor	0.17	0.11	0.26	0.39

Source: GAO analysis of NRC data.

Note: This table includes data on the 104 commercial nuclear reactors operating in the United States through the end of calendar year 2012. In February 2013, the owner of the Crystal River Nuclear Plant in Florida permanently shut down that plant's reactor; in May 2013, the owner of the Kewaunee Power Station in Wisconsin permanently shut down that plant's reactor; and in June 2013, the owner of the San Onofre Nuclear Generating Station in California permanently shut down that plant's two reactors. These actions reduced the number of operating commercial nuclear reactors in the United States from 104 to 100.

Differences in the amount of time reactors are under increased oversight. NRC officials stated that differences in the amount of time reactors are under increased oversight as a result of performance deficiencies— defined as being in column 2 or higher on NRC's action matrix—may contribute to differences in the number of nonescalated findings across regions, because increased oversight results in more scrutiny and more findings. However, according to 2001 through 2012 NRC data, the number of quarters that reactors in each region were under increased oversight were fairly consistent across regions when accounting for the number of reactors in each region. When comparing the number of nonescalated findings in each region to the number of quarters that reactors in each region were under increased oversight, and accounting for the regional differences in the number of reactors, we found that the results varied widely. Specifically, Region II identified the least number of nonescalated findings per quarter under increased oversight, and Region IV recorded the most, with more than 4 times as many nonescalated findings per quarter under increased oversight as Region II (see table 7). Furthermore, Region II had more reactors under increased oversight than Region IV for 36 of the 48 quarters we reviewed. Appendix VII summarizes NRC's performance assessments for each reactor from 2001 through 2012.

Differences in the age of reactors. NRC officials stated that differences in the age of reactors may contribute to the differences in the number of nonescalated findings across regions because nonescalated findings become more prevalent as reactors age. We analyzed the age of reactors across the regions and identified regional differences in their age. However, we found that reactors in Region II, which had the fewest nonescalated findings from 2000 through 2012, have been in commercial operation, on average, more years than, and have collectively been in commercial operation for almost twice as long as, reactors in Region IV, which had the most nonescalated findings over that period (see table 8).

Differences in resources available to plants. NRC officials stated that, when one licensee owns or manages multiple plants (collectively referred to as a licensee's "fleet"), benefits derived from plants' combined resources may contribute to the differences in the number of nonescalated findings across regions, because some regions have more plants in fleets than others. For example, according to NRC officials, licensees often require all plants that they manage or own to make improvements in response to findings made at only one of their plants, which could lead all the plants in this fleet to experience fewer findings over time.

Table 7. Nonescalated Findings and Amount of Time under Increased Oversight (per NRC's Action Matrix), Calendar Years 2001-2012

Nonescalated findings, quarters under increased oversight, and reactors	Region I (Northeast)	Region II (Southeast)	Region III (Midwest)	Region IV (West)
Number of nonescalated findings	2,315	1,778	2,982	3,059
Number of quarters reactors were under increased oversight[a]	274	375	305	233
Number of reactors	26	33	24	21
Average number of nonescalated findings per number of quarters under increased oversight per reactor	0.32	0.14	0.41	0.63

Source: GAO analysis of NRC data.

Notes: Based on NRC's assessment of licensee performance under the ROP, NRC places each of the licensee's reactors into one of five performance categories on its action matrix, which corresponds to graded, or increasing levels of oversight. The action matrix is NRC's formal method of determining how much additional oversight—mostly in the form of supplemental inspections—is required on the basis of the number and risk significance of performance indicators and inspection findings.

This table includes data on the 104 commercial nuclear reactors operating in the United States through the end of calendar year 2012. In February 2013, the owner of the Crystal River Nuclear Plant in Florida permanently shut down that plant's reactor; in May 2013, the owner of the Kewaunee Power Station in Wisconsin permanently shut down that plant's reactor; and in June 2013, the owner of the San Onofre Nuclear Generating Station in California permanently shut down that plant's two reactors. These actions reduced the number of operating commercial nuclear reactors in the United States from 104 to 100.

According to NRC officials, the number of inspection hours allocated to supplemental inspections conducted under increased oversight can vary by a factor of 75.

[a]The period of our review breaks down to 48 quarters. Not every reactor was under increased oversight for each quarter.

NRC officials told us that plants associated with fleets typically perform better because the licensee holds all plants to the same performance levels and provides resources and shares what would otherwise be proprietary information to make that happen. Some independent plants—those not part of a fleet—share resources by forming alliances with other plants, entering into management contracts with larger companies, or both. However, NRC officials stated that independent plants are limited in how much information they can share because some information is considered to be proprietary. We analyzed whether individual plants were associated with licensee fleets and found that being associated with a fleet may be a factor in the number of nonescalated findings a region identifies. For example, Region II, which had the fewest nonescalated findings from 2000 through 2012, had the highest percentage of its reactors located at plants associated with fleets. Conversely, Region IV, which had the most nonescalated findings during the same period, had the lowest percentage of its reactors located at plants associated with fleets (see table 9). However, because NRC has not conducted a comprehensive analysis of several potential factors, it is not clear to the extent this factor influences the differences in the number of findings across regions.

Table 8. Amount of Time Reactors per NRC Region Have Been in Commercial Operation

Years in commercial operation and reactors	Region I (Northeast)	Region II (Southeast)	Region III (Midwest)	Region IV (West)
Average number of years a reactor has been in commercial operation	33.9	33.4	34.2	28.3
Number of years reactors have collectively been in commercial operation	882	1103	822	594
Number of reactors	26	33	24	21

Source: GAO analysis of NRC data.

Note: This table includes data on the 104 commercial nuclear reactors operating in the United States through the end of calendar year 2012. In February 2013, the owner of the Crystal River Nuclear Plant in Florida permanently shut down that plant's reactor; in May 2013, the owner of the Kewaunee Power Station in Wisconsin permanently shut down that plant's reactor; and in June 2013, the owner of the San Onofre Nuclear Generating Station in California permanently shut down that plant's two reactors. These actions reduced the number of operating commercial nuclear reactors in the United States from 104 to 100.

In addition, NRC officials stated that the quality of reactor maintenance may contribute to the regional differences in the number of nonescalated findings. For example, the resident inspectors at a plant with a high number of findings in recent years told us that this was due, in part, to plant managers not fully addressing maintenance items for many years to save on costs. However, we could not analyze this factor because, according to NRC officials, the data are not readily available.

Table 9. Number of Reactors per NRC Region Associated with Licensee Fleets of Commercial Nuclear Power Plants

Region	Number of reactors associated with a licensee's fleet[a]	Total number of reactors in region	Percentage
I (Northeast)	16	26	62
II (Southeast)	31	33	94
III (Midwest)	17	24	71
IV (West)	5	21	24
Total	69	104	66

Source: GAO analysis of NRC data.

Note: This table includes data on the 104 commercial nuclear reactors operating in the United States through the end of calendar year 2012. In February 2013, the owner of the Crystal River Nuclear Plant in Florida permanently shut down that plant's reactor; in May 2013, the owner of the Kewaunee Power Station in Wisconsin permanently shut down that plant's reactor; and in June 2013, the owner of the San Onofre Nuclear Generating Station in California permanently shut down that plant's two reactors. These actions reduced the number of operating commercial nuclear reactors in the United States from 104 to 100.

[a]For this report, a "fleet" is defined as reactors across two or more plants that are owned or managed by the same licensee.

NRC Officials and Industry Representatives Have Raised Concerns about Differences in the Number of Findings Identified across NRC Regions

NRC officials and industry representatives have raised concerns that differences in the number of findings NRC regions identify may be due, in part, to differences in how NRC staff identify and resolve findings. NRC officials, including some resident inspectors in Region IV, told us that they believe Region IV identifies nonescalated findings more aggressively than other regional offices and that some regional branch chiefs differently interpret the criteria for assessing performance deficiencies and what constitutes more-than-minor findings. Some NRC officials also noted that the regions have flexibility to combine issues into a single finding or to split them into multiple findings, which may also account for some of the regional variation in the number of findings identified. These officials told us that, in their view, this autonomy and discretion is a good and desirable part of the ROP. Nonetheless, in 2008, NRC's IG reported that NRC was inconsistently resolving findings across its regional offices due, in part, to not having clear and comprehensive guidance needed to facilitate consistency.

NRC officials told us they survey industry annually as part of a ROP self-assessment process and have received feedback since the implementation of the ROP in 2000 about concerns from the appearance of differences in identifying and resolving findings across regions. They also stated that the Nuclear Energy Institute—the commercial nuclear trade organization—frequently raises these concerns with them. Plant managers we interviewed—several of whom have worked at plants across different regions—also raised concerns about these apparent differences. These plant managers told us that such differences could lead to several challenges, including focusing limited plant resources on a greater number of issues of very low risk significance (i.e., nonescalated findings).

NRC officials also told us that another concern that is sometimes raised about its findings by industry is that findings at different plants that appear to be similar sometimes result in different risk assessments. NRC officials provided the example of two findings at two different plants that appeared similar but that led to different assessments. Specifically, both findings dealt with problems with certain components of auxiliary feedwater pumps that rendered the pumps inoperable. Even though these findings appeared similar, they resulted in different risk significance levels—one green and one white—because they affected the safety of the two plants differently. NRC officials told us the process of assessing the risk significance of a finding is inherently complex and includes several factors, such as

- design of the plant, including availability of backup equipment and whether the site uses cooling towers or lakes or rivers for its ultimate source of cooling water;
- the effect that the finding had on the performance and availability of safety equipment;
- the amount of time the equipment was inoperable;
- criticality of the safety equipment associated with the finding; and
- historical reliability of the equipment in question for that plant design, among others.

Once NRC accounts for these factors, similar findings may result in different assessments because the risk that a finding poses is plant-specific, according to NRC officials.

NRC Regions Also Differed in the Cross-cutting Aspects They Assigned to Findings

The four NRC regions also differed in the number of cross-cutting aspects they assigned to findings and the percentage of findings assigned crosscutting aspects. For example, Regions III and IV had about twice as many findings assigned at least one cross-cutting aspect as Region II (see table 10).

Table 10. Number of Findings Assigned at Least One Cross-cutting Aspect per NRC Region, Calendar Years 2000-2012

	Cross-cutting aspect			
Region	Findings not assigned	Findings assigned at least one	Total	Percentage of findings assigned at least one
I (Northeast)	1,044	1,666	2,710	61
II (Southeast)	1,047	1,046	2,093	50
III (Midwest)	1,446	1,936	3,382	57
IV (West)	1,147	2,192	3,339	66
Total	**4,684**	**6,840**	**11,524**	**59**

Source: GAO analysis of NRC data.

Notes: If NRC officials determine that one of the cross-cutting areas—human performance, safety-conscious work environment, and problem identification and resolution—contributed to a finding, NRC assigns a cross-cutting aspect within that area to the finding. More than one cross-cutting aspect may be assigned to a finding.

NRC data did not include information about cross-cutting aspects for 397 findings.

This table includes data on the 104 commercial nuclear reactors operating in the United States through the end of calendar year 2012. In February 2013, the owner of the Crystal River Nuclear Plant in Florida permanently shut down that plant's reactor; in May 2013, the owner of the Kewaunee Power Station in Wisconsin permanently shut down that plant's reactor; and in June 2013, the owner of the San Onofre Nuclear Generating Station in California permanently shut down that plant's two reactors. These actions reduced the number of operating commercial nuclear reactors in the United States from 104 to 100.

Table 11 shows the types of cross-cutting aspects assigned to findings across regions.

Table 11. Types of Cross-cutting Aspects Assigned to Findings per NRC Region, Calendar Years 2000-2012

Region	Cross-cutting aspect: Safety-conscious work environment	Cross-cutting aspect: Human performance	Cross-cutting aspect: Problem identification and resolution	Percentage: Total	Percentage: total
I (Northeast)	3	878	835	1,716	25
II (Southeast)	4	637	415	1,056	15
III (Midwest)	3	1,254	693	1,950	28
IV (West)	4	1,389	856	2,249	32
Total	14	4,158	2,799	6,971	100

Source: GAO analysis of NRC data.

Notes: If NRC officials determine that one of the cross-cutting areas—human performance, safety-conscious work environment, and problem identification and resolution—contributed to a finding, NRC assigns a cross-cutting aspect within that area to the finding. More than one cross-cutting aspect may be assigned to a finding.

NRC data did not include information about cross-cutting aspects for 397 findings.

This table includes data on the 104 commercial nuclear reactors operating in the United States through the end of calendar year 2012. In February 2013, the owner of the Crystal River Nuclear Plant in Florida permanently shut down that plant's reactor; in May 2013, the owner of the Kewaunee Power Station in Wisconsin permanently shut down that plant's reactor; and in June 2013, the owner of the San Onofre Nuclear Generating Station in California permanently shut down that plant's two reactors. These actions reduced the number of operating commercial nuclear reactors in the United States from 104 to 100.

According to NRC guidance, identifying cross-cutting aspects is NRC's primary means of identifying trends. The guidance states that a crosscutting aspect is a performance characteristic of a finding that is the most significant causal factor of the performance deficiency. The guidance further states that the ROP was developed with the presumption that inspection findings with cross-cutting aspects would help them identify the plants with significant performance issues. Therefore, according to NRC officials, the greater the number of findings with cross-cutting aspects, the greater the chance that NRC could assign a substantive cross-cutting issue to the plant. Once NRC assigns a substantive cross-cutting issue to a plant, closing the issue requires additional efforts by the licensee and additional oversight from NRC. The number of plants assigned substantive cross-cutting issues at some time during calendar years 2001 through 2012 varied across the regions (see table 12). In Region II, less than half the plants were assigned a substantive cross-cutting issue, whereas a substantial majority of plants in other regions were assigned substantive cross-cutting issues. Appendix VIII summarizes the substantive cross-cutting issues assigned to plants from 2001 through 2012.

Table 12. Number of Nuclear Power Plants Assigned Substantive Cross-cutting Issues by NRC Region, Calendar Years 2001-2012

Region	Nuclear power plants assigned substantive cross-cutting issues	Total nuclear power plants	Percentage assigned substantive cross-cutting issues
I (Northeast)	12	17	71
II (Southeast)	7	18	39
III (Midwest)	15	16	94
IV (West)	12	14	86
Total	46	65	71

Source: GAO analysis of NRC data.

Notes: Every 6 months, NRC regional officials analyze all findings issued at each plant during the prior 12-month assessment period. In general, if a nuclear plant has four findings with the same cross-cutting aspect assigned during the previous year, and NRC is concerned about the licensee's progress in addressing them, NRC determines that the nuclear plant has what NRC calls a substantive cross-cutting issue.

This table includes data on the 65 commercial nuclear power plants operating in the United States through the end of calendar year 2012. In February 2013, the owner of the Crystal River Nuclear Plant in Florida permanently shut down that plant's reactor; in May 2013, the owner of the Kewaunee Power Station in Wisconsin permanently shut down that plant's reactor; and in June 2013, the owner of the San Onofre Nuclear Generating Station in California permanently shut down that plant's two reactors. These actions reduced the number of operating commercial nuclear power plants in the United States from 65 to 62.

According to NRC officials, because the criteria for closing a substantive cross-cutting issue involve professional judgment, determining whether to close one is somewhat subjective. The relevant regional office considers whether to close a substantive cross-cutting issue in its semiannual or annual assessment letters to licensees. Industry officials told us that, because NRC only considers closing substantive cross-cutting issues in these 6-month increments, it is difficult to close a substantive crosscutting issue in a timely fashion.

Number of Escalated Findings Had Fewer Differences across NRC Regions

The number of escalated findings had fewer differences across regional offices. As shown in table 13, each region had between 46 and 93 escalated findings from 2000 through 2012. Region III had more Severity Level III violations than the other regions, but otherwise, the number of escalated findings was largely consistent. Appendix IX lists the number of escalated findings identified at each plant from 2000 through 2012.

Table 13. Escalated Findings per NRC Region, Calendar Years 2000-2012

Escalated findings, plants, and reactors	Region I (Northeast)	Region II (Southeast)	Region III (Midwest)	Region IV (West)
Number of escalated findings	54	64	93	46
Number of plants	17	18	16	14
Number of reactors	26	33	24	21
Average number of escalated findings per reactor	2.1	1.9	3.9	2.2

Source: GAO analysis of NRC data.

Notes: The greater number of escalated findings in Region III over the other regions is due to the identification of more than three times as many Severity Level III violations.

This table includes data on the 104 commercial nuclear reactors operating in the United States through the end of calendar year 2012. In February 2013, the owner of the Crystal River Nuclear Plant in Florida permanently shut down that plant's reactor; in May 2013, the owner of the Kewaunee Power Station in Wisconsin permanently shut down that plant's reactor; and in June 2013, the owner of the San Onofre Nuclear Generating Station in California permanently shut down that plant's two reactors. These actions reduced the number of operating commercial nuclear reactors in the United States from 104 to 100.

NRC officials told us that escalated findings are identified and addressed more consistently than nonescalated findings because they receive a much higher level of review than do nonescalated findings. For example, according to NRC's enforcement policy, regional offices—subject to oversight by NRC headquarters—typically require licensees to provide a written response to an escalated finding that describes the reasons for the identified deficiency, actual and proposed corrective actions, and the expected date of compliance. Regional offices then conduct supplemental inspections for most escalated findings. Supplemental inspections expand the scope of baseline inspections to review, in part (1) the extent of the problem, (2) the sufficiency of the licensee's evaluation of the root cause of the problem, and (3) the licensee's proposed corrective actions. In addition, NRC officials told us that headquarters officials participate in meetings with regional officials and inspectors to ensure that an escalated finding is accurately assessed and that any proposed enforcement actions are justifiable. Moreover, according to NRC guidance, the Director of NRC's Office of Enforcement (or designee)— located in headquarters— has to approve any enforcement actions against licensees in response to escalated findings.

NRC Has Taken Some Steps to Reduce Differences across Regions Concerning Nonescalated Findings but Has Not Conducted a Comprehensive Analysis

NRC has taken some steps to reduce differences across the regions in the identification and resolution of nonescalated findings but has not conducted a comprehensive analysis of the differences to determine its consistency in handling findings or reasons for the differences.

A key effort NRC has undertaken to reduce differences in how the regions implement the ROP is the ROP Reliability Initiative, which NRC's regional offices initiated in 2009. According to NRC officials, the initiative was intended to explore how the different regional offices identify and assess inspection findings. Some NRC officials said that the initiative examines issues related to consistency, such as how to control the subjectivity in making distinctions between minor and more-than-minor performance deficiencies and violations. Each region took the lead on one of the following specific efforts:

- *Enhanced inspection resource sharing among regions*: This effort was a pilot program for sharing inspection resources across regional offices. Inspectors from different regions and resident inspectors from other plants participated on several types of team inspections. This initiative was led by Region IV and began in 2009.
- *Branch chief visits to other regions*: This effort included each region designating one branch chief to visit each of the other regions to identify best practices and implementation differences across the regions. This effort was led by Region II and began in 2009.
- *Periodic discussion of reliability topics*: This effort included quarterly videoconferences among regional officials to share insights on the

application of ROP inspection guidance and gain insight into different regional decision making. This effort was led by Region I and began in 2009.

- *ROP self-assessments of inspection report quality:*[35] This effort included a team of staff from all regions that selected a cross-section of inspection reports from each region to assess their conformance with applicable guidance, identify key differences between regions, and identify any weaknesses. This effort was led by Region III and concluded in 2012.

The ROP Reliability Initiative has not resulted in a formal report, and according to NRC officials, the sample of activities reviewed was not large enough to draw agencywide conclusions. In late 2011, NRC's regional offices decided to continue the ROP Reliability Initiative by applying the approach of the individual efforts to an in-depth review of a specific inspection area each fiscal year. For example, the first such review done in fiscal year 2012 resulted in an internal memorandum making 25 recommendations to provide additional guidance and training on limited aspects of NRC's efforts to inspect for weaknesses in licensees' corrective action programs. NRC officials told us that the overall initiative is still under way and that ongoing activities are intended to improve regions' consistency of actions adhering to the ROP guidance; however, agency officials said that, due to the ROP Enhancement Project (see below) and increased inspection activities following the Fukushima accident, the regional offices have not identified a specific inspection area to focus on for the fiscal year 2013 review.

NRC has also conducted other efforts aimed at improving the consistency of its oversight processes:

- *Assessment of regional enforcement.* According to NRC officials, in 2011, NRC established an initiative aimed at improving consistency in how regional offices resolve inspection findings through enforcement activities. This initiative was conducted to address findings of enforcement inconsistency in the 2008 report by the NRC IG.[36] Officials stated that NRC recently completed four regional assessments that were intended to improve consistency of interpreting minor and more-than-minor thresholds for violations. According to NRC officials, this initiative resulted in a final report in May 2013 recommending that regional enforcement staff conduct periodic audits of nonescalated enforcement actions.

- *"Minor" ROP realignment.* According to NRC officials, in 2006, NRC undertook an internal review of substantive cross-cutting issues across regions. The review resulted, in part, from complaints from industry that some plants were assigned cross-cutting aspects more frequently than others, especially in Region IV. NRC officials stated that Region IV led the review, which found that, while each region was following guidance, it also had its own culture and exercised some subjective judgment. Therefore, the regions had some differences in interpretations, especially in their approaches to determining what constitutes a more-than-minor performance deficiency. In response, according to NRC officials, NRC conducted additional training but made no key changes to its ROP procedures.

In addition, NRC officials said that several of the agency's formal ongoing reviews examine, to a limited extent, the consistency of its oversight processes:

- *ROP Enhancement Project.* In January 2013, as part of its annual self-assessment, NRC initiated the ROP Enhancement Project, which is intended as a long-term, strategic review of how it implements the ROP, among other things. NRC officials said that they expect this review to consider regional differences in the number of nonescalated findings, but according to an NRC status report, the assessment portion of the review has been delayed pending the outcome of an independent review of the ROP. NRC officials told us they expect procedural changes in response to the project to begin in 2014.
- *ROP realignment process.* In 2009, NRC implemented the ROP realignment process, a biennial review of its ROP baseline inspection procedures and inspection hours per procedure to ensure the most effective overall applications of inspection resources. NRC officials told us that this effort also touches on the consistency of how NRC implements its oversight of safety.
- *Self-assessments.* In 2001, NRC initiated annual self-assessments, which are to assess whether the ROP met its program goals and achieved its intended outcomes. NRC officials said that these self-assessments consider, to a limited extent, issues related to consistency, and according to NRC guidance, these self-assessments assess, in part, the extent to which NRC implements the ROP predictably.

However, the biennial ROP realignment process evaluates specific inspection procedures agencywide to ensure the most effective application of inspection resources, and according to NRC guidance, the self-assessments are not intended to audit the performance of the regions in implementing the ROP. Thus even with taking these steps to examine the consistency of its oversight processes, agency officials told us that NRC has not conducted a comprehensive analysis of the causes of the differences in the number of nonescalated findings identified across regions. Under federal standards for internal control, managers at the functional or activity level are to compare actual performance to planned or expected results throughout the organization and analyze significant differences.[37] Without conducting such an analysis, NRC (1) does not know whether its regional offices or individual inspectors are applying regulations and guidance consistently in handling findings; or (2) cannot identify other factors that may have led to the differences.

Moreover, NRC's 2008 to 2013 strategic plan states that the agency will establish and maintain stable and predictable regulatory programs and policies, and other NRC guidance states that the agency should use an objective, understandable, and predictable process to ensure that licensees fulfill their responsibility for ensuring the safe operation of commercial nuclear reactors. In particular, the agency's enforcement manual emphasizes that staff should ensure that oversight efforts are objective and consistent. NRC is aware of differences across regions in identifying and resolving the findings that result from its oversight processes. However, without a comprehensive analysis of the causes of the differences in the number of nonescalated findings across regions, NRC cannot ensure that oversight efforts are objective and consistent, as specified in its enforcement manual, or know whether its regional offices or individual inspectors are applying regulations and guidance consistently.

Records of Findings Are Difficult to Search and Assess

NRC maintains records of its identification of and response to both escalated and nonescalated findings; however, these records are difficult to search for several reasons, and limited our ability to assess the extent to which NRC followed its procedures and, thus, consistently identified and resolved findings. According to NRC's strategic plan, the agency's organizational values include openness in communication and decision making. Furthermore, NRC guidance states that the agency communicates ROP results to, in part, keep external stakeholders informed of licensee performance and to enhance confidence that NRC is accomplishing its mission. NRC's enforcement policy states that its mission includes promoting the transparency and openness of its enforcement program for all stakeholders.

In reviewing NRC's findings data since the ROP's inception in 2000 through 2012, we found that NRC has recorded approximately 12,000 findings—about 2 percent of which were escalated. For 95 percent of the findings data we reviewed—including numerous escalated findings—NRC did not include a case number for each finding, making it time-consuming and difficult to locate all documents associated with a single finding. According to NRC officials, although nonescalated findings each have a unique identifier, only the case number is searchable in NRC's databases. NRC officials told us that, as a matter of policy, they do not assign case numbers to nonescalated findings; only to escalated findings. However, our review of NRC findings data also found 38 instances of escalated findings that had not been assigned a case number, during this period.

The information NRC makes publicly available on its identification of and response to findings was also difficult to use. NRC communicates the results of much of its oversight process to the public through its ROP website. This website makes available inspection reports, assessment letters, and other general materials related to NRC's oversight process. NRC also provides a quarterly summary of every reactor's performance, consisting of its inspection findings, the color of each performance indicator, and the reactors' placement on the action matrix. However, to review this information, users must search a link for each plant on a quarterly basis. Since the ROP has been in place for more than 12 years, users may have to search dozens of links per plant and read many different documents to get complete information on individual findings. In addition, while descriptions in each document contain information about whether a finding was associated with a cross-cutting issue, the website itself does not provide information on those plants that had open substantive cross-cutting issues prior to the most recent assessment period. This information can be found on the website by linking to each plant's individual assessment letters. NRC program officials acknowledged that it is difficult for users of the website to determine which plants had substantive cross-cutting issues open in the past. Therefore, it is difficult to identify, over time, those that NRC views as having potentially more significant problems.

NRC also makes publicly available its official recordkeeping system, the Agencywide Documents Access and Management System (ADAMS). ADAMS contains more than 730,000 documents dating back to November 1999,[38] and several hundred new documents on NRC's reactor oversight processes—as well as other NRC activities—are added each day. However, we found ADAMS difficult to navigate. For example, using an inspection report as our starting point, we tried to retrace the steps in NRC's response to the findings in the report using ADAMS; however, our searches resulted in numerous unrelated documents as well as documents with ambiguous names that had to be opened to determine whether they were related to the findings. Further, we had no way of determining whether our searches returned a complete set of results and had identified all the documents related to the findings. NRC officials acknowledged that, while ADAMS can be used to conduct searches on specific words or other criteria, such searches may produce a large number of unrelated documents, in addition to relevant documents.

We asked NRC officials how we could track a finding through its resolution, and they told us that doing so is difficult because findings and enforcement actions are recorded in separate databases and that these databases are not linked. The officials told us that users with expertise in both databases' search capabilities could craft a data search that linked inspection and enforcement records. However, because external users cannot craft such a search, they cannot independently track NRC's identification and resolution of findings. NRC's Chairman reiterated the agency's support for transparency in prepared remarks before the annual NRC Regulatory Information Conference in March 2013, stating that the agency must have a continued commitment to openness and transparency and communicate with interested parties in ways they understand, and that clear, consistent communication is a principle of good regulation. However, because external users cannot independently track findings and their resolution, NRC cannot ensure that the public, Congress, and others are kept informed of licensee performance, as specified in guidance, to enhance confidence that NRC is accomplishing its mission and following its oversight processes.

NRC Uses Numerous Methods to Develop Lessons Learned for Improving Oversight, but Challenges Remain in Accessing This Information

NRC has both formal and informal methods for developing lessons learned to improve its oversight. Formal methods include agencywide programs and annual assessments. Informal methods include reaching out to peer groups and technical experts throughout the agency and accessing various agency databases. Although NRC guidance directs inspectors to use information in agency databases on past experiences to plan and conduct future inspection activities, inspectors face challenges accessing this information, which may limit their ability to use it.

NRC Has Formal Methods for Collecting and Analyzing Information to Develop Lessons Learned

NRC has ongoing, formal methods through which it collects and analyzes information to develop lessons learned for improving reactor oversight, including the following agencywide programs:

- *Operating Experience Program*, which captures information on past experience based on inspection findings, events, and documents from industry, international nuclear agencies, and others to inform agency activities, including reactor inspections.
- *Generic Issues Program*, which tracks the status and resolutions of certain issues involving public health and safety, among other things, that could affect multiple entities under NRC jurisdiction, such as multiple commercial nuclear plants.
- *ROP Feedback Program*, which enables staff to submit comments and suggestions for improving the ROP based on their experiences; a central coordinator then directs these comments and suggestions to appropriate internal stakeholders for their review.
- *Industry Trends Program*, which monitors trends in indicators of industry performance. NRC evaluates any adverse trends to determine whether regulatory action is appropriate, and reports statistically significant trends to Congress.
- *Monthly ROP working group meetings*, which, according to NRC officials, NRC hosts and include the Nuclear Energy Institute, other industry representatives, and members of the public to discuss operating experience, inspection and performance assessment topics, potential changes and questions regarding performance indicator guidance, and other topics.

Additionally, according to NRC documents, NRC identifies lessons learned during its annual self-assessments to determine whether the ROP met goals and during its biennial ROP realignment process. For example, the ROP realignment process assesses feedback from inspectors and other NRC officials submitted through the ROP Feedback Program.

NRC has also undertaken special initiatives, such as the Fukushima Near-Term Task Force, to develop lessons learned from the accident at the Fukushima Daiichi Nuclear Power Plant in Japan that it could apply in its oversight of U.S. commercial nuclear power reactors. NRC convened the task force in 2011 to review its processes and regulations and determine whether lessons learned from the accident could inform its oversight processes. The task force concluded that a sequence of events like the Fukushima accident is unlikely to occur in the United States and that the continued operation of the nation's commercial nuclear reactors does not pose an imminent risk to public health and safety. In addition, in 2011 the task force made 12 recommendations, such as for NRC to clarify its regulatory framework and to require licensees to reevaluate and upgrade seismic and flooding protection of reactors and related equipment, strengthen capabilities at all reactors to withstand loss of electrical power, and take other actions to better protect their plants for a low-probability high-impact event. NRC accepted these recommendations and is in the process of implementing them (see app. X).

To follow up on the task force's recommendations, in 2011, NRC established the Japan Lessons Learned Project Directorate—a group of over 20 NRC staff focused on implementing the lessons the task force identified. As part of this effort, NRC issued the following three orders to licensees in March 2012:

- improve capability to reliably and remotely monitor a wider range of conditions in the pools that contain spent fuel;[39]
- install vent systems for certain types of reactors to help prevent an excess buildup of hydrogen gas in the reactor buildings;
- enhance plants' capabilities to maintain or restore core cooling, containment, and spent fuel pool cooling in the case of external events, such as an earthquake, that exceeds plants' design capabilities.

NRC has also undertaken special initiatives in response to other unique events. For example, following the terrorist attacks on September 11, 2001, NRC issued orders to licensees to enhance security measures to protect their reactors against an increased terrorism threat. In addition, in 2002, after extensive degradation was discovered in the reactor vessel head at the Davis-Besse Nuclear Power Station in Ohio,[40] NRC established a lessons learned task force and implemented many of its recommendations, including inspection procedure changes, enhancing its training program, and enhancing aspects of program management of the ROP. In May 2004, we made several recommendations to ensure that NRC addresses weaknesses that contributed to the Davis-Besse incident, which NRC has since implemented.[41]

NRC Also Has Informal Methods for Collecting and Analyzing Information to Develop Lessons Learned

NRC officials told us they also have more informal methods of collecting and analyzing information to develop lessons learned for improving reactor oversight. These informal means include reaching out to peers by telephone or e-mail, reaching out to regional or headquarters technical experts by telephone or e-mail, and informal gatherings during peer group meetings. For example, several resident inspectors told us they typically rely on a small group of fellow resident inspectors—often within the same branch overseeing a small number of plants—that they call periodically for updates on issues about which they are known to have prior experience.

Inspectors told us that this type of outreach is especially important because it allows them to tap into other inspectors' different experiences performing inspections and unique engineering expertise, given that resident inspectors come from a variety of engineering backgrounds. They also told us that this form of outreach can have a large influence on an inspector's perspective about potential findings. Resident inspectors stated that they hold many informal discussions about lessons learned— including sharing findings and reviewing changes to guidance—during their semiannual conferences with peers from across their region and during periodic visits to plants in other regions. NRC uses e-mail to distribute updates agencywide to its guidance and other policy statements on how to respond to lessons learned from different plants; however, according to inspectors, individual inspectors are responsible for incorporating these lessons learned into their oversight activities. In addition, according to NRC officials, each region determines its policy for e-mailing lessons learned from across the region or lessons learned with generic safety implications.

Challenges Accessing Information on Past Experiences May Limit Inspectors' Ability to Benefit from It

Although NRC guidance directs inspectors to use information on past experiences to plan and conduct future inspection activities, inspectors reported facing challenges in accessing information on past experiences, which may limit their ability to benefit from it. Under NRC guidance, inspectors are to use information on past plant experiences to prepare for, conduct, and document inspection activities. According to several inspectors, this information can be important for helping them choose inspection samples. Some inspectors we interviewed also stated that complete access to lessons learned from past plant experience is important to help them understand how NRC identified and resolved similar findings at other plants. In addition, industry representatives have stated that information on past experiences is beneficial because the industry is mature and there is a robust body of information on past experiences to learn from.

However, several NRC officials told us that the Reactor Operating Experience (ROE) Gateway—the agency's internal website for the Operating Experience Program—and its search tools are not user-friendly. For example, they noted that the website provides links to more than a dozen search tools that all contain different information but does not explain how to use them. The officials said that providing instructions on how to use these tools would be helpful. Moreover, they told us that conducting a thorough search on a specific topic may require multiple keyword searches because different reactor equipment manufacturers use different terms to describe the same type of equipment. For example, one manufacturer of commercial nuclear reactors calls certain pumps Auxiliary Feedwater Pumps, while another manufacturer calls them Emergency Feedwater Pumps, although both serve the same function— removing heat from the reactor core when the plant experiences a complete loss of electrical power. As a result, inspectors and other agency users—including officials in the Operating Experience Program and technical specialists in the regional offices—have to conduct searches on several possible keywords in each of the various search tools. NRC officials told us the various terms used to describe the same equipment are not officially documented so staff must learn about them through their own experiences. For example, if an NRC official encountered a problem at a plant with a particular type of pump, using ROE Gateway to search for past experiences with this pump would require the official to know that this piece of equipment had more than one name and to search on both names in each of the search tools. At least one of the search tools on ROE Gateway now uses drop-down lists to eliminate the need to conduct multiple keyword searches, but the remaining search tools do not have this capability.

Given the challenges in navigating ROE Gateway, several resident inspectors told us they prefer contacting other inspectors informally or requesting assistance from NRC headquarters staff when trying to obtain information on past experience. They also told us they have developed informal "workarounds" to facilitate searches, such as creating and sharing slide presentations with one another on keyword searches for various common inquiries. We have previously reported that informal relationships could end once personnel move to their next assignments.[42] Thus an informal workaround cannot substitute for a user-friendly, searchable system. Without better search tools, inspectors could overly rely on information available through informal channels rather than the more comprehensive information located in agency databases. NRC guidance states that the efficient retrieval of information on past experiences is a basic requirement for meeting agencywide objectives and that such information is necessary to improve safety assessments and NRC decisions. Further, NRC officials noted that inspectors have varied backgrounds—that is, some may have expertise in mechanical engineering, while others have expertise in electrical engineering— therefore they depend on NRC's network of information to assess most issues and determine their importance. Past inspection results and related information are crucial parts of that network. NRC officials have acknowledged the difficulty of using ROE Gateway and noted at their most recent annual meeting in March 2013 that NRC has struggled to make information about past experiences accessible and distribute this information efficiently.

Several resident inspectors also told us that e-mail updates on past experiences could be managed more effectively. For example, once resident inspectors receive an e-mail, it is their responsibility to incorporate the information on past experiences at their plant. However, these resident inspectors noted that not all e-mail updates apply to all plants; for example, an update may apply to a certain type of reactor that is not in use at all plants. Some resident inspectors stated that one way to improve this system would be to prioritize the e-mail notices and indicate the extent to which they are relevant to the inspectors. In addition, some resident inspectors noted that nonurgent e-mail updates should be compiled, whereas now they are included alongside all e-mail updates.

Resident inspectors play an essential role in ensuring that reactors are operating safely because they have the opportunity and authority to, at any time, physically inspect plant equipment and operations, review plant records, and verify the accuracy of licensee-reported quantitative measures or indicators of plant performance. NRC officials told us that, because of their essential role in NRC's oversight, inspectors need complete and efficient access to all the information necessary to do their jobs effectively. In addition, information on past plant performance and oversight actions helps inspectors know where to begin looking for potential issues of concern, and NRC guidance states that inspectors should consider this information in preparing for, conducting, and documenting inspection activities. For example, prior to the reactor damage found at the Davis-Besse plant in 2002, inspectors did not regularly focus on identifying reactor vessel head corrosion; however, as a result of this incident, NRC inspectors now focus on this issue regularly.

CONCLUSION

NRC oversees the safety of commercial nuclear power reactors using established, ongoing processes. NRC's oversight has taken on even greater importance as many commercial reactors in the United States are operating under renewed licenses that will allow them to operate beyond their initial 40-year operating periods. NRC is aware of differences across its regions in the identification and resolution of findings that result from its oversight processes and has taken some steps to address them, particularly in the differences of nonescalated findings. However, while NRC has undertaken these steps, NRC has not conducted a comprehensive analysis of these differences consistent with federal standards of internal control. Without a comprehensive analysis of the differences in the number of nonescalated findings across regions and their causes, NRC cannot ensure that oversight efforts are objective and consistent, as specified in its enforcement manual, or know whether its regional offices or individual inspectors are not applying regulations and guidance consistently.

Furthermore, NRC policy states that the agency's mission includes promoting transparency and openness for all stakeholders. As part of such transparency, NRC makes publicly available its official recordkeeping system, which contains hundreds of thousands of documents on NRC's reactor oversight processes—as well as other NRC activities. However, we found this system difficult to navigate, and external users cannot use it to independently track findings, all documents related to the findings, and the findings' resolution. As a result, NRC cannot ensure that the public, Congress, and others are kept informed of licensee performance, as specified in agency guidance, to enhance public confidence that NRC is accomplishing its mission and following its oversight processes.

Information about reactor operators' past experiences and NRC's responses to them also plays a key role in NRC's oversight processes. Although NRC guidance directs inspectors to use information on past experiences to plan and conduct future inspection activities, inspectors reported facing challenges in accessing information on past experiences, which may limit their ability to benefit from this information. Given the challenges in navigating ROE Gateway, several resident inspectors reported contacting other inspectors informally or requesting assistance from NRC headquarters staff when trying to obtain information on past experience. Without better search tools, inspectors could overly rely on information available through informal channels. NRC guidance states that the efficient retrieval of information on past experiences is a basic requirement for meeting agencywide objectives and that such information is necessary to improve safety assessments and NRC decisions.

RECOMMENDATIONS FOR EXECUTIVE ACTION

To improve NRC's oversight processes, we are making three recommendations to the NRC Commissioners:

- To better meet its goal of implementing objective and consistent oversight, direct agency managers to conduct a comprehensive analysis of the causes of the differences in the identification and resolution of findings.
- To improve transparency and better enable the public, Congress, and others to independently track findings, all documents related to the findings, and the findings' resolution, direct the agency to either modify NRC's publicly available recordkeeping system to do so or develop a publicly accessible tool that does so.
- To help NRC staff more efficiently use past experiences in their oversight activities, direct agency officials to evaluate the challenges inspectors face in retrieving all relevant information on plant performance and previous oversight activities and improve its systems accordingly to address these challenges.

Agency Comments

We provided a draft copy of this report to the Executive Director for Operations of NRC for review and comment. In its written comments, NRC generally agreed with our findings and recommendations. NRC also provided technical comments, which we incorporated, as appropriate.

To address our first recommendation, NRC stated that it will revisit its initiatives with respect to implementation of the ROP and the nonescalated enforcement process to identify potential enhancements. For our second recommendation, NRC stated that it will identify ways to improve transparency and track documents related to inspection findings through improved tools to facilitate public access to inspection information. To address our third recommendation, NRC stated that it will make plant performance and oversight information more readily searchable and available to inspection staff and other NRC personnel.

In addition, in its comment letter, concerning regional differences, NRC stated that, although it believes adequate internal controls to ensure alignment between regions exist through program office oversight and audits, among other things, it agrees to seek enhancements in this area. NRC raised concerns about repeated references in the draft report to the use of professional judgment by NRC staff members in the ROP and Traditional Enforcement Process, stating that these references imply a high degree of subjectivity in the agency's implementation of its oversight processes and that its use of professional judgment is excessive or inconsistent. NRC stated that it believes its use of professional judgment is limited and controlled through its guidance. As stated in the report, we agree that NRC's oversight processes are largely prescribed in guidance.

Frank Rusco
Director, Natural Resources and Environment

Appendix I: Objectives, Scope, and Methodology

Our objectives were to (1) describe how the Nuclear Regulatory Commission (NRC) implements its processes for overseeing the safety of commercial nuclear power reactors; (2) evaluate the extent to which NRC consistently identifies and resolves findings through these processes; and (3) describe NRC's methods for developing lessons learned to improve its oversight and challenges, if any, NRC faces in doing so.

To describe how NRC implements its processes for overseeing the safety of commercial nuclear power reactors, we reviewed relevant federal laws, regulations, and guidance, including NRC's *Enforcement Policy*, *Enforcement Manual*, and *Inspection Manual*, among many other guidance documents. We also reviewed previous GAO and NRC Inspector General reports that identified safety-related challenges or made recommendations for improving NRC's oversight of reactor safety, and we reviewed documentation of NRC efforts to update its oversight processes. We interviewed NRC headquarters officials from the Offices of Nuclear Reactor Regulation, Enforcement, Nuclear Regulatory Research, and the Inspector General. We also conducted semistructured interviews with officials from all four NRC regions, as well as resident inspectors and plant representatives during site visits to five nuclear power plants (discussed further below). During these semistructured interviews, we asked NRC officials and plant representatives about (1) the implementation and consistency of NRC's oversight activities, (2) the extent to which NRC's oversight includes lessons learned from operating experience and safety-related findings, (3) the level of communication between NRC and plant representatives, (4) NRC actions taken to identify and resolve specific safety-related findings, and (5) site-specific considerations related to safety. In addition, we interviewed representatives from the Nuclear Energy Institute, the Union of Concerned Scientists, academic specialists, and other groups and individuals to discuss their views and gain additional insight into NRC's oversight of reactor safety.

We made site visits to a nonprobability sample of five nuclear power plants. During these site visits, we held separate semistructured interviews with NRC resident inspectors and plant representatives, including licensee managers and plant operators, to obtain in-depth information about their experiences with NRC's oversight process. We also toured several of the sites with the resident inspectors and observed areas and equipment related to previous findings, in addition to current issues of concern being assessed for potential findings. Because this was a nonprobability sample, the information we gathered from these site visits is not generalizable to all U.S. nuclear power plants but provides important illustrative information for understanding the oversight of reactors at the selected plants. To select these sites, we applied the following criteria to capture a variety of characteristics: whether NRC's assessment of the plant's safety significance has changed in the past 2 years, and whether, collectively, they were located in multiple NRC regions, operated by different licensees, and included different reactor types. In Region I, we visited the Calvert Cliffs Nuclear Power Plant in Maryland. In Region III, we visited the Byron Station in Illinois and Palisades Nuclear Plant in Michigan. In Region IV, we visited the Cooper Nuclear Station in Nebraska and Wolf Creek Generating Station in Kansas.

To evaluate the extent to which NRC consistently identifies and resolves findings through its processes for overseeing reactor safety, in addition to reviewing NRC guidance and interviewing NRC officials (as described above), we analyzed the nonsecurity findings and violations for all plants from the beginning of NRC's Reactor Oversight Process in 2000 through 2012.[1] We obtained these data from NRC's Reactor Program System database. We also analyzed separate NRC data on (1) the number of inspection hours from 2000 through 2012, (2) descriptive information about licensees, (3) substantive cross-cutting issues assigned to each plant from 2001 through 2012, and (4) the quarterly action matrix assessment for each reactor from 2001 through 2012. In order to assess the reliability of the data we analyzed, we reviewed database documentation, interviewed NRC officials familiar with the data, and conducted electronic tests of the data, looking for missing values, outliers, or other anomalies. We determined that the data were sufficiently reliable for our purposes. We also searched documents in NRC's publicly available, web-based Agencywide Documents Access and Management System (ADAMS) database, which is the official recordkeeping system for all NRC documents.

To describe NRC's methods for developing lessons learned to improve its oversight and challenges, if any, NRC faces in doing so, in addition to reviewing NRC guidance and interviewing NRC officials (as described above), we observed NRC headquarters officials demonstrating the various computer-based tools they use for communicating and providing access to information on past experiences and lessons learned. We also observed resident inspectors, during our site visits to nuclear power plants, demonstrating how they access agency information on past experiences and lessons learned. In addition, we observed meetings where NRC staff updated the Commission and the public on the status of the development and implementation of lessons learned from the Fukushima disaster, among other safety-related concerns.

We conducted this performance audit from May 2012 to September 2013 in accordance with generally accepted government auditing standards. Those standards require that we plan and perform the audit to obtain sufficient, appropriate evidence to provide a reasonable basis for our findings and conclusions based on our audit objectives. We believe that the evidence obtained provides a reasonable basis for our findings and conclusions based on our audit objectives.

APPENDIX II: SUMMARY INFORMATION ABOUT OPERATING NUCLEAR POWER REACTORS IN THE UNITED STATES

This appendix provides summary information about the 104 commercial nuclear power reactors at 65 power plants for which we analyzed NRC data and that were operating in the United States through the end of calendar year 2012. For each reactor, table 14 lists the NRC region, state, date commercial operation began, and licensee.

Table 14. Summary Information about Operating Commercial Nuclear Power Reactors

Reactor	State	Date commercial operation began	Licensee
NRC Region I			
Beaver Valley Power Station, Unit 1	Pennsylvania	Oct. 1976	First Energy Nuclear Operating Co.
Beaver Valley Power Station, Unit 2	Pennsylvania	Nov.1987	First Energy Nuclear Operating Co.
Calvert Cliffs Nuclear Power Plant, Unit 1	Maryland	May 1975	Calvert Cliffs Nuclear Power Plant Inc.
Calvert Cliffs Nuclear Power Plant, Unit 2	Maryland	Apr. 1977	Calvert Cliffs Nuclear Power Plant Inc.
Hope Creek Generating Station, Unit 1	New Jersey	Dec. 1986	PSEG Nuclear, LLC
Indian Point Nuclear Generating, Unit 2	New York	Aug. 1974	Entergy Nuclear Operations, Inc.
Indian Point Nuclear Generating, Unit 3	New York	Aug. 1976	Entergy Nuclear Operations, Inc.
James A. FitzPatrick Nuclear Power Plant	New York	July 1975	Entergy Nuclear Operations, Inc.
Limerick Generating Station, Unit 1	Pennsylvania	Feb. 1986	Exelon Generation Co., LLC

Reactor	State	Date commercial operation began	Licensee
Limerick Generating Station, Unit 2	Pennsylvania	Jan. 1990	Exelon Generation Co., LLC
Millstone Power Station, Unit 2	Connecticut	Dec. 1975	Dominion Nuclear Connecticut, Inc.
Millstone Power Station, Unit 3	Connecticut	Apr. 1986	Dominion Nuclear Connecticut, Inc.
Nine Mile Point Nuclear Station, Unit 1	New York	Dec. 1969	Nine Mile Point Nuclear Station, LLC
Nine Mile Point Nuclear Station, Unit 2	New York	Mar. 1988	Nine Mile Point Nuclear Station, LLC
Oyster Creek Nuclear Generating Station, Unit 1	New Jersey	Dec. 1969	Exelon Generation Co., LLC
Peach Bottom Atomic Power Station, Unit 2	Pennsylvania	July 1974	Exelon Generation Co., LLC
Peach Bottom Atomic Power Station, Unit 3	Pennsylvania	Dec. 1974	Exelon Generation Co., LLC
Pilgrim Nuclear Power Station	Massachusetts	Dec .1972	Entergy Nuclear Operations, Inc.
R.E. Ginna Nuclear Power Plant	New York	July 1970	R.E. Ginna Nuclear Power Plant, LLC
Salem Nuclear Generating Station, Unit 1	New Jersey	June 1977	PSEG Nuclear, LLC
Salem Nuclear Generating Station, Unit 2	New Jersey	Oct. 1981	PSEG Nuclear, LLC
Seabrook Station, Unit 1	New Hampshire	Aug. 1990	FPL Energy Seabrook, LLC
Susquehanna Steam Electric Station, Unit 1	Pennsylvania	June 1983	PPL Susquehanna, LLC
Susquehanna Steam Electric Station, Unit 2	Pennsylvania	Feb. 1985	PPL Susquehanna, LLC
Three Mile Island Nuclear Station, Unit 1	Pennsylvania	Sept. 1974	Exelon Generation Co., LLC
Vermont Yankee Nuclear Power Plant, Unit 1	Vermont	Nov. 1972	Entergy Nuclear Operations, Inc.
Browns Ferry Nuclear Plant, Unit 1	Alabama	Aug. 1974	Tennessee Valley Authority
Browns Ferry Nuclear Plant, Unit 2	Alabama	Mar. 1975	Tennessee Valley Authority
Browns Ferry Nuclear Plant, Unit 3	Alabama	Mar. 1977	Tennessee Valley Authority

Table 14. (Continued)

Reactor	State	Date commercial operation began	Licensee
Brunswick Steam Electric Plant, Unit 1	North Carolina	Mar. 1977	Carolina Power & Light Co.
Brunswick Steam Electric Plant, Unit 2	North Carolina	Nov. 1975	Carolina Power & Light Co.
Catawba Nuclear Station, Unit 1	South Carolina	June 1985	Duke Energy Carolinas, LLC
Catawba Nuclear Station, Unit 2	South Carolina	Aug. 1986	Duke Energy Carolinas, LLC
Crystal River Nuclear Generating Plant, Unit 3	Florida	Mar. 1977	Florida Power Corp.
Edwin I. Hatch Nuclear Plant, Unit 1	Georgia	Dec. 1975	Southern Nuclear Operating Co.
Edwin I. Hatch Nuclear Plant, Unit 2	Georgia	Sept. 1979	Southern Nuclear Operating Co.
H. B. Robinson Steam Electric Plant, Unit 2	South Carolina	Mar. 1971	Carolina Power & Light Co.
Joseph M. Farley Nuclear Plant, Unit 1	Alabama	Dec. 1977	Southern Nuclear Operating Co.
Joseph M. Farley Nuclear Plant, Unit 2	Alabama	July 1981	Southern Nuclear Operating Co.
McGuire Nuclear Station, Unit 1	North Carolina	Dec. 1981	Duke Energy Carolinas, LLC
McGuire Nuclear Station, Unit 2	North Carolina	Mar. 1984	Duke Energy Carolinas, LLC
North Anna Power Station, Unit 1	Virginia	June 1978	Virginia Electric & Power Co.
North Anna Power Station, Unit 2	Virginia	Dec. 1980	Virginia Electric & Power Co.
Oconee Nuclear Station, Unit 1	South Carolina	July 1973	Duke Energy Carolinas, LLC
Oconee Nuclear Station, Unit 2	South Carolina	Sept. 1974	Duke Energy Carolinas, LLC
Oconee Nuclear Station, Unit 3	South Carolina	Dec. 1974	Duke Energy Carolinas, LLC
Sequoyah Nuclear Plant, Unit 1	Tennessee	July 1981	Tennessee Valley Authority
Sequoyah Nuclear Plant, Unit 2	Tennessee	June 1982	Tennessee Valley Authority
Shearon Harris Nuclear Power Plant, Unit 1	North Carolina	May 1987	Carolina Power & Light Co.

Reactor	State	Date commercial operation began	Licensee
St. Lucie Plant, Unit 1	Florida	Dec. 1976	Florida Power & Light Co.
St. Lucie Plant, Unit 2	Florida	Aug. 1983	Florida Power & Light Co.
Surry Nuclear Power Station, Unit 1	Virginia	Dec. 1972	Virginia Electric & Power Co.
Surry Nuclear Power Station, Unit 2	Virginia	May 1973	Virginia Electric & Power Co.
Turkey Point Nuclear Generating, Unit 3	Florida	Dec. 1972	Florida Power & Light Co.
Turkey Point Nuclear Generating, Unit 4	Florida	Sept. 1973	Florida Power & Light Co.
Virgil C. Summer Nuclear Station, Unit 1	South Carolina	Jan. 1984	South Carolina Electric & Gas Co.
Vogtle Electric Generating Plant, Unit 1	Georgia	June 1987	Southern Nuclear Operating Co.
Vogtle Electric Generating Plant, Unit 2	Georgia	May 1989	Southern Nuclear Operating Co.
Watts Bar Nuclear Plant, Unit 1	Tennessee	May 1996	Tennessee Valley Authority
Braidwood Station, Unit 1	Illinois	July 1988	Exelon Generation Co., LLC
Braidwood Station, Unit 2	Illinois	Oct. 1988	Exelon Generation Co., LLC
Byron Station, Unit 1	Illinois	Sept. 1985	Exelon Generation Co., LLC
Byron Station, Unit 2	Illinois	Aug. 1987	Exelon Generation Co., LLC
Clinton Power Station, Unit 1	Illinois	Nov. 1987	Exelon Generation Co., LLC
Davis-Besse Nuclear Power Station, Unit 1	Ohio	July 1978	First Energy Nuclear Operating Co.
Donald C. Cook Nuclear Power Plant, Unit 1	Michigan	Aug. 1975	Indiana Michigan Power Co.
Donald C. Cook Nuclear Power Plant, Unit 2	Michigan	July 1978	Indiana Michigan Power Co.
Dresden Nuclear Power Station, Unit 2	Illinois	June 1970	Exelon Generation Co., LLC
Dresden Nuclear Power Station, Unit 3	Illinois	Nov. 1971	Exelon Generation Co., LLC
Duane Arnold Energy Center	Iowa	Feb. 1975	FPL Energy Duane Arnold, LLC

Table 14. (Continued)

Reactor	State	Date commercial operation began	Licensee
Fermi, Unit 2	Michigan	Jan. 1988	The Detroit Edison Co.
Kewaunee Power Station	Wisconsin	June 1974	Dominion Energy Kewaunee, Inc.
LaSalle County Station, Unit 1	Illinois	Jan. 1984	Exelon Generation Co., LLC
LaSalle County Station, Unit 2	Illinois	Oct. 1984	Exelon Generation Co., LLC
Monticello Nuclear Generating Plant, Unit 1	Minnesota	June 1971	Northern States Power Company
Palisades Nuclear Plant	Michigan	Dec. 1971	Entergy Nuclear Operations, Inc.
Perry Nuclear Power Plant, Unit 1	Ohio	Nov. 1987	First Energy Nuclear Operating Co.
Point Beach Nuclear Plant, Unit 1	Wisconsin	Dec. 1970	FPL Energy Duane Arnold, LLC
Point Beach Nuclear Plant, Unit 2	Wisconsin	Oct. 1972	FPL Energy Duane Arnold, LLC
Prairie Island Nuclear Generating Plant, Unit 1	Minnesota	Dec. 1973	Northern States Power Co. Minnesota
Prairie Island Nuclear Generating Plant, Unit 2	Minnesota	Dec. 1974	Northern States Power Co. Minnesota
Quad Cities Nuclear Power Station, Unit 1	Illinois	Feb. 1973	Exelon Generation Co., LLC
Quad Cities Nuclear Power Station, Unit 2	Illinois	Mar. 1973	Exelon Generation Co., LLC
NRC Region IV			
Arkansas Nuclear One, Unit 1	Arkansas	Dec. 1974	Entergy Nuclear Operations, Inc.
Arkansas Nuclear One, Unit 2	Arkansas	Mar. 1980	Entergy Nuclear Operations, Inc.
Callaway Plant	Missouri	Dec. 1984	Union Electric Co.
Columbia Generating Station, Unit 2	Washington	Dec. 1984	Energy Northwest
Comanche Peak Steam Electric Station, Unit 1	Texas	Aug. 1990	Luminant Generation Co., LLC

Reactor	State	Date commercial operation began	Licensee
Comanche Peak Steam Electric Station, Unit 2	Texas	Aug. 1993	Luminant Generation Co., LLC
Cooper Nuclear Station	Nebraska	July 1974	Nebraska Public Power District
Diablo Canyon Nuclear Power Plant, Unit 1	California	May 1985	Pacific Gas & Electric Co.
Diablo Canyon Nuclear Power Plant, Unit 2	California	Mar. 1986	Pacific Gas & Electric Co.
Fort Calhoun Station, Unit 1	Nebraska	Sept. 1973	Omaha Public Power District
Grand Gulf Nuclear Station, Unit 1	Mississippi	July 1985	Entergy Nuclear Operations, Inc.
Palo Verde Nuclear Generating Station, Unit 1	Arizona	Jan. 1986	Arizona Public Service Company
Palo Verde Nuclear Generating Station, Unit 2	Arizona	Sept. 1986	Arizona Public Service Company
Palo Verde Nuclear Generating Station, Unit 3	Arizona	Jan. 1988	Arizona Public Service Company
River Bend Station, Unit 1	Louisiana	June 1986	Entergy Nuclear Operations, Inc.
San Onofre Nuclear Generating Station, Unit 2	California	Aug. 1983	Southern California Edison Co.
San Onofre Nuclear Generating Station, Unit 3	California	Apr. 1984	Southern California Edison Co.
South Texas Project, Unit 1	Texas	Aug. 1988	STP Nuclear Operating Co.
South Texas Project, Unit 2	Texas	June 1989	STP Nuclear Operating Co.
Waterford Steam Electric Station, Unit 3	Louisiana	Sept. 1985	Entergy Nuclear Operations, Inc.
Wolf Creek Generating Station, Unit 1	Kansas	Sept. 1985	Wolf Creek Nuclear Operating Corp.

Source: NRC.

Note: This table includes data on the 104 commercial nuclear power reactors operating in the United States through the end of calendar year 2012. In February 2013, the owner of the Crystal River Nuclear Plant in Florida permanently shut down that site's reactor; in May 2013, the owner of the Kewaunee Power Station in Wisconsin permanently shut down that site's reactor; and in June 2013, the owner of the San Onofre Nuclear Generating Station in California permanently shut down that site's two reactors. These actions reduced the number of operating commercial nuclear power reactors in the United States from 104 to 100.

Appendix III: NRC Safety Cornerstones and Related Performance Indicators

Initiating Events cornerstone– This cornerstone focuses on operations and events at a nuclear plant that could lead to a possible accident, if plant safety systems did not intervene. These events could include equipment failures leading to a plant shutdown, shutdowns with unexpected complications, or large changes in the plant's power output. (In this table, "scram" refers to automatic and manual reactor shutdowns.)	
Performance indicator	**Objective**
Unplanned scrams per 7,000 critical hours	This performance indicator monitors the number of unplanned scrams. It measures the rate of scrams per year of operation at power and provides an indication of initiating event frequency. The value of 7,000 hours is used because it represents 1 year of reactor operation at 80 percent capacity.
Unplanned scrams with complications	This performance indicator monitors the subset of unplanned scrams while critical that require additional operator actions beyond that of the "normal" scram. Such events or conditions have the potential to present additional challenges to the plant operations staff and therefore, may be more risk-significant than uncomplicated scrams.
Unplanned power changes per 7,000 critical hours	This performance indicator monitors the number of unplanned power changes (excludingscrams) that could have, under other plant conditions, challenged safety functions. It may provide leading indication of risk-significant events but is not itself risk-significant. The indicator measures the number of plant power changes for a typical year of operation at power.
Mitigating Systems cornerstone– This cornerstone measures the function of safety systems designed to prevent an accident or reduce the consequences of a possible accident. The equipment is checked by periodic testing and through actual performance.	
Performance indicator	**Objective**
Mitigating system performance index – emergency alternating current power systems	This performance indicator monitors the readiness of the emergency alternating current power systems to perform their safety functions in response to off-normal events or accidents.
Mitigating system performance index – high pressure injection systems	This performance indicator monitors the readiness of the high pressure injection systemsto perform their safety functions in response to off-normal events or accidents.

Mitigating system performance index –heat removal systems	This performance indicator monitors the readiness of the heat removal systems to perform their safety functions in response to off-normal events or accidents.
Mitigating system performance index – residual heat removal systems	This performance indicator monitors the readiness of the residual heat removal systems to perform their safety functions in response to off-normal events or accidents.
Mitigating system performance index – cooling water systems	This performance indicator monitors the readiness of the cooling water support systems to perform their safety functions in response to off-normal events or accidents.
Safety system functional failures	This performance indicator monitors events or conditions that alone prevented, or could have prevented, the fulfillment of the safety function of structures or systems that are needed to: shut down the reactor and maintain it in a safe shutdown condition; remove residual heat; control the release of radioactive material; or mitigate the consequences ofan accident.
Barrier Integrity cornerstone – This cornerstone focuses on the physical barriers between the highly radioactive materials in fuel within the reactor and the public and the environment outside the plant. These barriers are the sealed rods containing the fuel pellets, the heavy steel reactor vessel and associated piping, and the reinforced concrete containment building surrounding the reactor. The integrity of the fuel rods, the vessel, and the piping is continuously checked for leakage, while the ability of the containment to prevent leakage is measured on a regular basis.	
Performance indicator	**Objective**
Reactor coolant system specific activity	This performance indicator monitors the integrity of the fuel cladding, the first of the threebarriers to prevent the release of fission products. It measures the radioactivity in the reactor coolant system as an indication of functionality of the cladding.
Reactor coolant system leakage	This performance indicator monitors the integrity of the reactor coolant system pressure boundary, the second of the three barriers to prevent the release of fission products. It measures reactor coolant system identified leakage as a percentage of the technical specification allowable identified leakage to provide an indication of reactor coolant system integrity.

(Continued)

Emergency Preparedness cornerstone–This cornerstone measures the effectiveness of plant staff in carrying out emergency plansfor responding to a possible accident. Such emergency plans are tested every 2 years during emergency exercises involving the plant staff and local, state, and, in some cases, federal agencies.	
Performance indicator	**Objective**
Drill/exercise performance	This performance indicator monitors timely and accurate licensee performance in drills and exercises when presented with opportunities for classification of emergencies, notification of off-site authorities, and development of protective action recommendations. It is the ratio, in percent, of timely and accurate performance of those actions to total opportunities.
Emergency response organization drill participation	This performance indicator tracks the participation of emergency response organization members assigned to fill key positions in performance enhancing experiences, and through linkage to the drill/exercise performance indicator ensures that the risk-significant aspects of classification, notification, and protective action recommendations development are evaluated and included in the performance indicator process. This indicator measures the percentage of emergency response organization members assigned to fill key positions who have participated recently in performance-enhancing experiences such as drills, exercises, or in an actual event.
Alert and notification system reliability	This performance indicator monitors the reliability of the off-site alert and notification system, a critical link for alerting and notifying the public of the need to take protective actions. It provides the percentage of the sirens that are capable of performing their safety function based on regularly scheduled tests.
Occupational Radiation Safety cornerstone– NRC regulations set a limit on radiation doses received by plant workers, and this cornerstone monitors the effectiveness of the plant's program to control and minimize those doses.	
Performance indicator	**Objective**
Occupational exposure control effectiveness	This performance indicator monitors the control of access to and work activities within radiologically-significant areas of the plant and occurrences involving degradation or failure of radiation safety barriers that result in readily-identifiable unintended dose.

Public Radiation Safety cornerstone– This cornerstone measures the procedures and systems designed to minimize radioactive releases from a nuclear plant during normal operations and to keep those releases within federal limits.	
Performance indicator	**Objective**
"Radiological effluent technical specifications" / "offsite dose calculation manual" radiological effluent occurrence	This performance indicator assesses the performance of the program to control radiological effluent occurrences, which are radiological releases that exceed pre-set limits for liquid and gaseous releases.
Physical Protection cornerstone[a]– Nuclear plants are required to have well-trained security personnel and a variety of protectivesystems to guard vital plant equipment, as well as programs to assure that employees are constantly fit for duty through drug and alcohol testing. This cornerstone measures the effectiveness of the security and fitness-for-duty programs.	

Sources: GAO analysis of NRC and Nuclear Energy Institute information.

[a]The Physical Protection cornerstone consists of physical security issues. Our review focuses on NRC's oversight of the safe operation of reactors; therefore the physical security of plants and related performance indicators are outside the scope of this review.

APPENDIX IV: NRC SAFETY CORNERSTONES AND RELATED INSPECTION PROCEDURES

This appendix provides summary information about NRC's inspection procedures. Table 15 lists the safety cornerstone, title, and objective for each of the 36 safety-related (nonsecurity) baseline inspection procedures. Table 16 lists the title and objective for each of the 3 supplemental inspection procedures, and table 17 lists the title and objective for each of the 3 reactive inspection procedures.

Table 15. NRC Safety Cornerstones and Related Baseline Inspection Procedures

Initiating Events cornerstone– This cornerstone focuses on operations and events at a nuclear plant that could lead to a possible accident, if plant safety systems did not intervene. These events could include equipment failures leading to a plant shutdown, shutdowns with unexpected complications, or large changes in the plant's power output.

(Continued)

Mitigating Systems cornerstone– This cornerstone measures the function of safety systems designed to prevent an accident or reduce the consequences of a possible accident. The equipment is checked by periodic testing and through actual performance. **Barrier Integrity cornerstone**– This cornerstone focuses on the physical barriers between the highly radioactive materials in fuel within the reactor and the public and the environment outside the plant. These barriers are the sealed rods containing the fuel pellets, the heavy steel reactor vessel and associated piping, and the reinforced concrete containment building surrounding the reactor. The integrity of the fuel rods, the vessel, and the piping is continuously checked for leakage, while the ability of the containment to prevent leakage is measured on a regular basis.		
Inspection procedure	**Title**	**Objective**
71111.01	Adverse weather protection	This inspection focuses on verifying that the design features and implementation of the licensee's procedures protect mitigating systems from adverse weather effects. Adverse weather includes events such as high winds, hurricanes, electrical storms, tornadoes, and extreme high or low temperatures.
71111.04	Equipment alignment	This inspection focuses on verifying equipment alignment, identifying any discrepancies that impact the function(s) of the system, and verifying that the licensee has properly identified and resolved equipment alignment problems.
71111.05AQ	Fire Protection (Annual/Quarterly)	This inspection focuses on evaluating the licensee's fire protection program, includingassessing the performance of the fire brigade and verifying the adequacy of its fire detection and suppression capability and controls for combustibles and ignition sources within the plant, among other things.
71111.05T	Fire Protection (Triennial)[a]	This inspection focuses on evaluating the design, operational status, and material condition of the licensee's fire protection program. In addition, inspection teams verify that certain mitigating strategies are feasible in light of operator training, maintenance of necessary equipment, and any plant modifications.

Inspection procedure	Title	Objective
		This inspection is designed to complement Inspection Procedure 71111.05AQ.
71111.05XT	Fire Protection – NFPA 805 (Triennial)[b]	This inspection focuses on evaluating the design, installation, operational status, testing, and material condition of the licensee's fire protection program for plants, as well as the plant's capability to meet requirements of the NRC-approved fire protection program, and goals, objectives, and criteria for fire safety of National Fire Protection Association (NFPA) Standard 805. In addition, inspection teams verify that certain mitigating strategies are feasible in light of operator training, maintenance of necessary equipment, and any plant modifications. This inspection is designed to complement Inspection Procedure 71111.05AQ.
71111.06	Flood Protection Measures	This inspection focuses on verifying that the licensee's flooding mitigation plans and equipment are consistent with the licensee's design requirements and the risk analysis assumptions.
71111.07	Heat Sink Performance	This inspection focuses on verifying that (1) potential heat exchanger deficiencies which could mask degraded performance are identified; (2) potential heat sink performance problems that have the potential to increase risk are identified; and (3) the licensee has adequately resolved them.
71111.08	Inservice Inspection Activities	This inspection focuses on assessing the effectiveness of the licensee's program for monitoring degradation of reactor coolant system boundary, risk-significant piping system boundaries, and the containment boundary.
71111.11	Licensed Operator Requalification Program and Licensed Operator Performance	This inspection focuses on evaluating licensed operator performance during facility-administered requalification examinations, other examinations, facility training exercises, and during selected

(Continued)

Inspection procedure	Title	Objective
		evolutions conducted in the actual plant/main control room. This inspection also reviews the licensee's ability to identify and correct problems associated with licensed operator performance.
71111.12	Maintenance Effectiveness	This inspection focuses on supplementing performance indicators by providing for independent oversight of licensee maintenance effectiveness.
71111.13	Maintenance Risk Assessments and Emergent Work Control	This inspection focuses on verifying the performance and adequacy of risk assessments for planned or emergent maintenance activities, among other activities.
71111.15	Operability Determinations and Functionality Assessments	This inspection focuses on reviewing the operability or functionality assessments for certain components or systems to ensure that they remain capable of performing their design functions.
71111.17T	Evaluations of Changes, Tests and Experiments and Permanent Plant Modifications	This inspection focuses on verifying that modifications to the plant have been adequately implemented, and that procedures and other documentation affected by the modifications have been adequately updated.
71111.18	Plant Modifications	This inspection focuses on verifying that modifications to the plant have not affected the safety functions of important safety systems, have not degraded certain structures, systems, and components, and have not placed the plant in an unsafe condition.
71111.19	Post-Maintenance Testing	This inspection focuses on verifying that post-maintenance test procedures and test activities are adequate to verify system operability, and functional capability.
71111.20	Refueling and Other Outage Activities	This inspection focuses on evaluating licensee outage activities, including verifying that the licensee considered risk in developing outage schedules and adhered to certain risk-reduction methodologies.

Inspection procedure	Title	Objective
71111.21	Component Design Bases Inspection	This inspection focuses on obtaining reasonable assurance that risk-significant structures, systems, and components can adequately perform their design basis function.
71111.22	Surveillance Testing	This inspection focuses on verifying that surveillance testing of risk-significant structures, systems, and components are capable of performing their intended safetyfunctions and assessing their operational readiness.
Emergency Preparedness cornerstone–This cornerstone measures the effectiveness of plant staff in carrying out emergency plansfor responding to a possible accident. Such emergency plans are tested every 2 years during emergency exercises involving the plant staff and local, state, and, in some cases, federal agencies.		
Inspection procedure	**Title**	**Description**
71114.01	Exercise Evaluation	This inspection focuses on evaluating the adequacy of the licensee's conduct of eachplant's biennial exercise, which each plant is required to complete to demonstrate their capability to adequately perform key skills in certain functional areas to protect public health and safety in the event of a radiological emergency.
71114.02	Alert and Notification System Testing	This inspection focuses on evaluating the licensee's compliance with the testing and maintenance requirements for alert and notification systems.
71114.03	Emergency Response Organization Staffing and Augmentation System	This inspection focuses on evaluating whether the licensee's on-shift and augmentation staffing levels meet emergency response commitments.
71114.04	Emergency Action Level and Emergency Plan Changes	This inspection focuses on assessing whether any changes to the licensee's emergency action level scheme have decreased the effectiveness of its emergency plan, and verifying that other emergency plan changes did not decrease its effectiveness.

(Continued)

Inspection procedure	Title	Objective
71114.05	Maintenance of Emergency Preparedness	This inspection focuses on evaluating the efficacy of licensee efforts to maintain their emergency preparedness programs.
71114.06	Drill Evaluation	This inspection focuses on evaluating the licensee's assessment of performance to identify weaknesses in emergency preparedness in selected drills and training evolutions and use of its corrective action program to correct these weaknesses.
71114.07	Exercise Evaluation – Hostile Action Event	This inspection focuses on evaluating the adequacy of the licensee's ability to implement mitigative measures in response to a simulated attack at the plant, coordinate required actions to successfully respond to and mitigate plant damage, as well as its capability to assess performance in order to identify and correct weaknesses.
Occupational Radiation Safety cornerstone– NRC regulations set a limit on radiation doses received by plant workers, and this cornerstone monitors the effectiveness of the plant's program to control and minimize those doses. **Public Radiation Safety cornerstone**– This cornerstone measures the procedures and systems designed to minimize radioactive releases from a nuclear plant during normal operations and to keep those releases within federal limits.		
Inspection procedure	**Title**	**Description**
71124.01	Radiological Hazard Assessment and Exposure Controls	This inspection focuses on assessing licensee performance in assessing the radiological hazards in the workplace and in implementing of appropriate radiation monitoring and exposure control measures. In addition, this inspection focuses on verifying that the licensee is properly identifying and reporting performance indicators for the Occupational Radiation Safety cornerstone.
71124.02	Occupational ALARA Planning and Controls	This inspection focuses on assessing licensee performance in maintaining radiation exposures as low as reasonably achievable (ALARA). This inspection will determine whether the licensee's ALARA program is effective.

Inspection procedure	Title	Objective
71124.03	In-Plant Airborne Radioactivity Control and Mitigation	This inspection focuses on verifying that in-plant airborne concentrations of radioactivity are as low as reasonably achievable to the extent necessary to validate plant operations and to verify that the practices and use of respiratory protection devices on site do not pose an undue risk to the wearer.
71124.04	Occupational Dose Assessment	This inspection focuses on determining the accuracy and operability of a licensee's personal radiation monitoring equipment and the accuracy and effectiveness of its methods for calculating and monitoring radiological exposure to workers.
71124.05	Radiation Monitoring Instrumentation	This inspection focuses on verifying that the licensee is ensuring the accuracy and operability of radiation monitoring instruments that are used to monitor areas, materials, and workers to ensure a radiologically safe work environment, among other things.
71124.06	Radioactive Gaseous and Liquid Effluent Treatment	This inspection focuses on ensuring that gaseous and liquid effluent processing systems are maintained so that radiological discharges are properly mitigated, monitored, and evaluated with regard to public exposure.
71124.07	Radiological Environmental Monitoring Program	This inspection focuses on verifying that the licensee quantifies the impact of radioactive effluent releases to the environment, that it sufficiently validates the integrity of its program that monitors radioactive gaseous and liquid effluent release, and that it implements this program consistently.
71124.08	Radioactive Solid Waste Processing and Radioactive Material Handling, Storage, and Transportation	This inspection focuses on verifying the effectiveness of the licensee's programs for processing, handling, storage, and transportation of radioactive material.

(Continued)

Inspection procedure	Title	Objective
Physical Protection cornerstone[c]– Nuclear plants are required to have well-trained security personnel and a variety of protective systems to guard vital plant equipment, as well as programs to assure that employees are constantly fit for duty through drug and alcohol testing. This cornerstone measures the effectiveness of the security and fitness-for-duty programs.		
Other baseline procedures		
Inspectionprocedure	**Title**	**Description**
71151	Performance Indicator Verification	This inspection focuses on performing a periodic review of performance indicator data to determine their accuracy and completeness.
71152	Problem Identification and Resolution	This inspection focuses on (1) providing an early warning of potential performance issues that could result in crossing thresholds in the action matrix of NRC's Reactor Oversight Process, (2) helping gauge NRC's response when these thresholds are crossed, and (3) providing insights into whether licensees have established a safety-conscious work environment, among other things.
71153	Follow-up of Events and Notices of Enforcement Discretion	This inspection focuses on evaluating licensee events and degraded conditions for plant status and mitigating actions to help determine the need for additional inspections.

Source: GAO analysis of NRC information.

[a]Traditionally, commercial nuclear power plants have followed a deterministic fire safety approach where plant operators are required to ensure that at least one system of electric cables and equipment is available to safely shut down a reactor if a fire occurs.

[b]In 2004 NRC issued a regulation permitting plants to voluntarily transition from a deterministic fire safety approach to risk-informed fire protection requirements. Under this approach, a licensee adopts performance goals, objectives, and criteria for fire safety that are defined by the fire protection standard—NFPA-805—issued by the National Fire Protection Association (NFPA), an international nonprofit organization that develops, publishes, and disseminates fire prevention and safety standards.

[c]This review focuses on NRC's oversight of the safe operation of reactors; therefore the physical security of plants (Physical Protection cornerstone) is outside the scope of this review.

Table 16. NRC Supplemental Inspection Procedures

Inspection procedure	Title	Description
95001	Supplemental Inspection for One or Two White Inputs in a Strategic Performance Area	This inspection focuses on (1) providing assurance that the root causes and contributing causes of risk-significant performance issues are understood, (2) that the extent of condition and extent of cause of risk-significant performance issues areidentified, and (3) that the licensee's corrective actions for risk-significant performance issues are sufficient to address them.
95002	Supplemental Inspection for One Degraded Cornerstone or any Three White Inputs in a Strategic Performance Area	This inspection focuses on (1) providing assurance that the cause(s) of risk-significant performance issues are understood, (2) independently assessing that the extent of the condition and extent of cause of risk-significant performance issues areidentified, (3) independently determining if safety culture components caused or contributed to these performance issues, and (4) providing assurance that the licensee's corrective actions are sufficient to prevent recurrence.
95003	Supplemental Inspection for Repetitive Degraded Cornerstones, Multiple Degraded Cornerstones, Multiple Yellow Inputs or One Red Input	This inspection focuses on providing the NRC additional information to be used in deciding whether the continued operation of the plant is acceptable and whether additional regulatory actions are necessary to stop the decline of a plant's performance. In addition, it focuses on providing an independent assessment of the extent of risk-significant issues to determine whether an unacceptable margin of safety or security exists, among other things.

Source: GAO analysis of NRC information.

Table 17. NRC Reactive Inspection Procedures

Inspection procedure	Title	Description
93812	Special Inspection	This inspection focuses on assessing a significant operational event and its causes with a special inspection team, including collecting, analyzing, and documenting information and evidence about the event, and evaluating the adequacy of the licensee's response to the event. The team conducting the inspection is composed of technical experts from the region in which the event took place. The special inspection team reports directly tothe appropriate regional administrator.
93800	Augmented Inspection Team	This inspection focuses on assessing a significant operational event and its causes with an augmented inspection team, including collecting, analyzing, and documenting information and evidence about the event, and evaluating the adequacy of the licensee's response to the event. It is similar to an inspection performed by a special inspection team except that the team conducting an augmented inspection consists of technical experts from the region in which the event took place and is augmented by personnel from headquarters or other regions or by contractors. The augmented inspection team reports directly to the appropriate regional administrator.
	Incident Investigation Team[a]	This inspection focuses on assessing a significant operational event and its causeswithin an incident investigation team, including collecting, analyzing, and documenting information and evidence about the event. The team conducting the inspection consists of technical experts who, to the extentpracticable, do not have, and have not had, previous significant involvement with licensing and inspection activities at the affected plant. The incident investigation team reports directly to the Executive Director for Operations and is independent ofregional and headquarters office management.

Source: GAO analysis of NRC information.

Note: A significant operational event is any radiological, safeguards, or other safety-related operational event at an NRC-licensed facility that poses an actual or a potential hazard to public health and safety, property, or the environment. A significant operational event also may be referred to as "an event" or "an incident."

[a]An incident investigation does not have an inspection procedure, according to an NRC official. Instead, it is described in NRC Management Directive 8.3, *NRC Incident Investigation Program*.

APPENDIX V: EXAMPLES OF NRC FINDINGS AND VIOLATIONS

This appendix provides summary information about select examples of NRC findings and violations. Table 18 provides a brief description, nuclear power plant location, and date recorded for select examples of NRC findings illustrating different risk significance levels. Table 19 provides a brief description, nuclear power plant location, and date recorded for select examples of NRC violations illustrating different severity levels.

Table 18. Select Examples of NRC Findings Illustrating Different Risk Significance Levels

Risk significance	Description	Nuclear power plant	Date recorded
Green finding[a]	Licensee did not verify the impact that High-Energy Line Breaks in the turbine building could have on safety-related electrical equipment. Determined to be of very low safety significance because it was a design deficiency confirmed not to result in a loss of operability. Entered in corrective action plan.	Indian Point Nuclear Generating	Nov. 2012
	Licensee failed to establish and perform adequate preventive maintenance on a certain transformer.	Limerick Generating Station	Sept. 2012
White finding[b]	Failure to establish and maintain emergency diesel generator maintenance procedures as required by regulation.	North Anna Power Station	Dec. 2011
	Flange on diesel generator failed, causing significant oil leak, and was required to be shut down.	Byron Station	Feb. 2011
Yellow finding[c]	Licensee failed to provide adequate oversight of contractor during maintenance of a startup transformer. Contractor replaced wires without necessary insulation, causing the wires to touch during operation, which caused a loss of off-site power to equipment.	Wolf Creek Generating Station	Jan. 2012

Table 18. (Continued)

Risk significance	Description	Nuclear power plant	Date recorded
	Failure to maintain external flood procedures—procedures that would protect intake structures and auxiliary building during external floods. The procedure to stack sandbags at certain height over floodgates deemed insufficient.	Fort Calhoun Station	June 2010
Red finding[d]	Failure to ensure that design requirements for electrical power distribution system were properly implemented and maintained. Led to catastrophic fire that affected the plant's ability to shut down safely.	Fort Calhoun Station	Apr. 2012
	Failure to restore a heat removal system to operating status within the required time.	Browns Ferry Nuclear Plant	Oct. 2010

Source: GAO analysis of NRC information.

[a]Green findings are denoted as very low safety significance; they indicate that licensee performance is acceptable and cornerstone objectives are fully met with nominal risk and deviation.

[b]White findings are denoted as low to moderate safety significance; they indicate that licensee performance is acceptable, but outside the nominal risk range. Cornerstone objectives are met with minimal reduction in safety margin.

[c]Yellow findings are denoted as substantial safety significance; they indicate a decline in licensee performance that is still acceptable with cornerstone objectives met, but with significant reduction in safety margin.

[d]Red findings are denoted as high safety significance; they indicate a decline in licensee performance that is associated with an unacceptable loss of safety margin. Sufficient safety margin still exists to prevent undue risk to public health and safety.

Table 19. Select Examples of NRC Violations Illustrating Different Severity Levels

Severity Level	Description	Nuclear power plant	Date recorded
Severity Level I[a]	Licensee submitted incomplete and inaccurate information related to completed corrective actions. Specifically, licensee did not remove accumulated boron deposits in certain areas.	Davis-Besse Nuclear Power Station	Aug. 2002

Severity Level	Description	Nuclear power plant	Date recorded
Severity Level II[b]	Failure to determine cause of having boric acid on reactor head and taking steps to preclude repetition.	Davis-Besse Nuclear Power Station	Aug. 2002
	Licensee did not select a former employee to a competitive position due, in part, to the employee's engagement in protected activities. The Atomic Safety and Licensing Board upheld NRC finding that licensee discriminated against its former employee.	Watts Bar Nuclear Plant	June 2001
Severity Level III[c]	A senior reactor operator performed licensed activities while license was inactive and was on temporary medical hold.	San Onofre Nuclear Generating Station	Apr. 2011
	Failure to provide complete and accurate information to NRC which impacted a licensing decision.	Donald C. Cook Nuclear Power Plant	Oct. 2006
Severity Level IV[d]	Failure to submit Licensee Event Report per 10 C.F.R. 50.73(a)(2)(vii).	Braidwood Station	Dec. 2011
Failure to obtain license amendment prior to implementing a proposed change to auxiliary feedwater system (as required by 10 C.F.R. 50.59).		Braidwood Station	Sept. 2011

Source: GAO analysis of NRC information.

[a]Severity Level I violations are violations that resulted in or could have resulted in serious safety consequences (e.g., violations that created the substantial potential for serious safety consequences or violations that involved systems failing when actually called on to prevent or mitigate a serious safety event).

[b]Severity Level II violations are violations that resulted in or could have resulted in significant safety consequences (e.g., violations that created the potential for substantial safety consequences or violations that involved systems not being capable, for an extended period, of preventing or mitigating a serious safety event).

[c]Severity Level III violations are violations that resulted in or could have resulted in moderate safety consequences (e.g., violations that created a potential for moderate safety consequences or violations that involved systems not being capable, for a relatively short period, of preventing or mitigating a serious safety event).

[d]Severity Level IV violations are violations that are less serious, but are of more than minor concern, that resulted in no or relatively inappreciable potential safety consequences (e.g., violations that created the potential of more than minor safety consequences).

APPENDIX VI: NRC NONESCALATED FINDINGS AT OPERATING NUCLEAR POWER PLANTS IN THE UNITED STATES

This appendix provides performance information about the 104 commercial nuclear power reactors at 65 power plants for which we analyzed NRC data and that were operating in the United States through the end of calendar year 2012. Table 20 lists the number of nonescalated findings identified at each plant from 2000 through 2012.

Table 20. NRC Nonescalated Findings (Green Inspection Findings and Severity Level IV Violations) per Operating Commercial Nuclear Power Plant, Calendar Years 2000-2012

Plant[a]	2000	2001	2002	2003	2004	2005	2006	2007	2008	2009	2010	2011	2012	Total
NRC Region I														
Beaver Valley Power Station	11	14	10	10	8	9	7	9	7	5	5	9	9	113
Calvert Cliffs Nuclear Power Plant	6	18	4	6	19	7	8	13	12	15	13	9	10	140
Hope Creek Generating Station	13	18	24	19	25	17	13	18	4	10	9	3	3	176
Indian Point Nuclear Generating[b]	27	57	24	34	31	24	30	34	27	26	26	20	20	380
James A. FitzPatrick Nuclear Power Plant	28	12	7	8	5	8	2	8	8	10	8	6	5	115
Limerick Generating Station	8	15	11	16	8	4	4	4	8	8	5	9	10	110
Millstone Power Station	17	20	13	8	12	17	17	13	14	12	13	8	9	173
Nine Mile Point Nuclear Station	8	16	9	15	19	8	4	10	13	12	7	8	11	140

Planta	2000	2001	2002	2003	2004	2005	2006	2007	2008	2009	2010	2011	2012	Total
Oyster Creek Nuclear Generating Station	10	11	12	7	15	13	16	13	9	14	15	10	8	153
Peach Bottom Atomic Power Station	10	14	13	20	7	9	7	9	12	13	4	8	5	131
Pilgrim Nuclear Power Station	10	7	4	9	6	7	5	9	10	9	8	17	7	108
R.E. Ginna Nuclear Power Plant	7	4	11	14	9	10	12	7	10	11	5	5	6	111
Salem Nuclear Generating Station	7	17	14	26	25	25	13	10	16	6	4	11	5	179
Seabrook Station	5	10	6	10	9	8	10	6	3	1	3	7	7	85
Susquehanna Steam Electric Station	20	11	17	11	14	7	11	11	11	8	17	19	18	175
Three Mile Island Nuclear Station	7	19	8	7	14	13	9	10	7	7	9	4	12	126
Vermont Yankee Nuclear Power Plant	9	11	9	9	15	3	5	6	12	5	8	5	6	103
Total (NRC Region I)	203	274	196	229	241	189	173	190	183	172	159	158	151	2,518
NRC Region II														
Browns Ferry Nuclear Plant[c]	5	6	5	4	10	8	10	19	7	14	18	15	14	135
Brunswick Steam Electric Plant	3	4	4	3	6	9	7	7	9	16	6	9	10	93
Catawba Nuclear Station	9	10	6	7	5	10	11	14	3	5	6	4	8	98
Crystal River Nuclear Generating Plant	5	5	4	5	5	6	2	7	5	6	7	3	0	60
Joseph M. Farley Nuclear Plant	5	8	5	5	4	11	6	5	6	14	11	10	6	96
Edwin I. Hatch Nuclear Plant	3	11	8	4	3	4	6	6	4	8	6	8	16	87

Table 20. (Continued)

Plant[a]	2000	2001	2002	2003	2004	2005	2006	2007	2008	2009	2010	2011	2012	Total
McGuire Nuclear Station	3	5	9	8	17	11	14	7	8	14	9	6	8	119
Plant[a]	2000	2001	2002	2003	2004	2005	2006	2007	2008	2009	2010	2011	2012	Total
North Anna Power Station	6	4	4	5	9	12	10	8	7	10	19	11	10	115
Oconee Nuclear Station	18	16	12	13	11	21	9	9	10	8	12	18	6	163
H.B. Robinson Steam Electric Plant	1	4	1	4	4	4	3	3	5	6	18	7	12	72
St. Lucie Plant	2	13	7	6	10	10	9	10	7	12	12	5	14	117
Sequoyah Nuclear Plant	14	11	16	9	7	4	8	8	6	8	8	11	13	123
Shearon Harris Nuclear Power Plant	8	5	5	5	5	6	4	7	6	5	11	15	8	90
Surry Nuclear Power Station	4	6	5	7	3	7	5	13	8	12	6	8	8	92
Turkey Point Nuclear Generating	4	6	2	9	13	9	8	13	8	9	9	9	7	106
Virgil C. Summer Nuclear Station	9	13	9	10	10	9	9	5	8	3	5	12	12	114
Vogtle Electric Generating Plant	3	4	8	6	8	3	3	7	5	7	6	6	9	75
Watts Bar Nuclear Plant	5	4	13	10	8	8	8	10	9	7	22	13	13	130
Total (NRC Region II)	107	135	123	120	138	152	132	158	121	164	191	170	174	1,885
NRC Region III														
Braidwood Station	5	12	12	6	5	8	9	13	14	17	22	33	22	178
Byron Station	6	14	10	17	14	20	9	13	19	10	15	22	13	182
Clinton Power Station	12	11	10	5	16	10	6	13	10	8	17	22	9	149
Davis-Besse Nuclear Power Station	11	5	16	25	52	13	14	15	8	9	13	14	6	201

Plant[a]	2000	2001	2002	2003	2004	2005	2006	2007	2008	2009	2010	2011	2012	Total
Donald C. Cook Nuclear Power Plant	14	14	27	25	8	17	14	11	9	6	5	3	15	168
Dresden Nuclear Power Station	18	20	19	22	26	13	22	17	10	22	19	23	10	241
Duane Arnold Energy Center	2	1	7	22	12	20	13	21	8	13	18	11	14	162
Fermi	7	6	6	12	14	22	14	24	16	7	10	11	11	160
Kewaunee Power Station	17	12	9	17	24	25	25	39	8	23	19	20	18	256
Lasalle County Station	6	9	13	7	12	21	6	18	5	5	7	16	11	136
Monticello Nuclear Generating Plant	3	17	10	12	13	11	9	9	17	12	10	17	14	154
Palisades Nuclear Plant	4	23	13	13	10	8	28	25	27	14	17	31	13	226
Perry Nuclear Power Plant	5	7	10	19	16	61	16	16	26	26	18	20	16	256
Point Beach Nuclear Plant	11	15	10	30	25	27	25	28	34	16	21	13	21	276
Prairie Island Nuclear Generating Plant	10	5	6	5	10	16	13	10	14	27	25	24	21	186
Quad Cities Nuclear Power Station	35	7	21	15	14	11	32	18	14	16	7	15	12	217
Total (NRC Region III)	166	178	199	252	271	303	255	290	239	231	243	295	226	3,148
NRC Region IV														
Arkansas Nuclear One	5	13	9	17	23	22	11	27	21	25	25	19	17	234
Callaway Plant	15	17	11	15	17	17	22	19	19	18	16	28	11	225
Columbia Generating Station	10	12	5	21	17	10	9	17	13	17	11	18	23	183
Comanche Peak Steam Electric Station	7	6	13	3	10	11	11	7	19	15	16	28	11	157
Cooper Nuclear Station	27	32	23	29	23	34	20	24	15	31	20	39	46	363
Plant[a]	2000	2001	2002	2003	2004	2005	2006	2007	2008	2009	2010	2011	2012	Total

Table 20. (Continued)

Plant[a]	2000	2001	2002	2003	2004	2005	2006	2007	2008	2009	2010	2011	2012	Total
Diablo Canyon Nuclear Power Plant	11	11	14	22	25	24	14	8	15	17	37	16	14	228
Fort Calhoun Station	11	7	18	12	15	21	18	18	19	13	23	15	29	219
Grand Gulf Nuclear Station	4	15	9	6	10	12	14	29	28	21	18	26	27	219
Palo Verde Nuclear Generating Station	4	16	3	4	49	32	39	43	22	29	14	18	26	299
River Bend Station	23	19	6	17	18	7	22	26	17	10	18	22	35	240
San Onofre Nuclear Generating Station	19	5	9	10	16	18	12	22	32	33	44	22	13	255
South Texas Project	11	14	7	15	13	14	7	10	14	10	10	17	9	151
Waterford Steam Electric Station	13	5	14	15	17	16	7	16	14	18	7	27	17	186
Wolf Creek Generating Station	6	5	3	7	9	7	15	27	43	44	38	38	24	266
Total (NRC Region IV)	166	177	144	193	262	245	221	293	291	301	297	333	302	3,225
Total (all regions)	642	764	662	794	912	889	781	931	834	868	890	956	853	10,776

Source: GAO analysis of NRC data.

Notes: This table includes data on the 104 commercial nuclear power reactors operating in the United States through the end of calendar year 2012. In February 2013, the owner of the Crystal River Nuclear Plant in Florida permanently shut down that site's reactor; in May 2013, the owner of the Kewaunee Power Station in Wisconsin permanently shut down that site's reactor; and in June 2013, the owner of the San Onofre Nuclear Generating Station in California permanently shut down that site's two reactors. These actions reduced the number of operating commercial nuclear power reactors in the United States from 104 to 100.

Green inspection findings are denoted as very low safety significance; they indicate that licensee performance is acceptable and cornerstone objectives are fully met with nominal risk and deviation.

Severity Level IV violations are violations that are less serious, but are of more than minor concern, that resulted in no or relatively inappreciable potential safety consequences (e.g., violations that created the potential of more than minor safety consequences).

[a]NRC reports these data by nuclear power plant as opposed to by individual reactor. Oftentimes, there are 2 or 3 reactors located at each plant. Therefore, data for all 104 reactors are included here, but at the plant level.

[b]Although NRC tracks findings at Indian Point Units 2 and 3 separately, they are physically located at the same plant.

[c]One of the Browns Ferry reactors (Unit 1) was shut down from March 1985 until May 2007 to address performance and management issues.

APPENDIX VII: SUMMARY OF NRC PERFORMANCE ASSESSMENTS FOR NUCLEAR POWER REACTORS OPERATING FROM 2001 THROUGH 2012

This appendix provides performance information about the 104 commercial nuclear power reactors at 65 power plants for which we analyzed NRC data and that were operating in the United States through the end of calendar year 2012. Based on NRC's assessment of licensee performance under the ROP, NRC places each of the licensee's reactors into one of five performance categories on its action matrix, which corresponds to increased levels of oversight. The action matrix is NRC's formal method of determining how much additional oversight—mostly in the form of supplemental inspections—is required on the basis of the number and risk significance of performance indicators and inspection findings. The level of oversight reported here corresponds to the highest level the plant received during the year, even if it was only a portion of the year. Thus, if a plant was placed into a different category each quarter, the highest category in which it was placed is reported here. (See table 21.)

Table 21. Highest Level of Oversight Applied by NRC to Operating Commercial Nuclear Power Reactors for One or More Quarters of Each Calendar Year, 2001-2012

Key:
1. Action matrix column 1: Licensee response
2. Action matrix column 2: Regulatory response
3. Action matrix column 3: Degraded cornerstone
4. Action matrix column 4: Multiple/repetitive degraded cornerstone
5. Action matrix column 5: Unacceptable performance
6. Removed from action matrix: IMC 0350 process

	Highest level of oversight (action matrixcolumn assigned) during at least some portion of the year											
Reactor	2001	2002	2003	2004	2005	2006	2007	2008	2009	2010	2011	2012
NRC Region I												
Beaver Valley Power Station, Unit 1	1	2	2	2	1	2	2	1	1	1	1	2
Beaver Valley Power Station, Unit 2	1	2	2	2	1	2	2	1	1	1	1	2
Calvert Cliffs Nuclear Power Plant, Unit 1	3	3	2	1	1	2	2	1	2	1	1	2
Calvert Cliffs Nuclear Power Plant, Unit 2	1	2	2	2	2	1	1	1	2	2	2	2
Hope Creek Generating Station, Unit 1	1	1	1	2	2	1	1	1	1	1	1	3
Indian Point Nuclear Generating, Unit 2	4	4	3	2	2	2	1	1	1	1	1	1
Indian Point Nuclear Generating, Unit 3	1	1	2	1	1	1	2	1	2	2	1	1
James A. FitzPatrick Nuclear Power Plant	2	1	1	1	1	1	1	1	1	1	1	2
Limerick Generating Station, Unit 1	2	2	1	1	1	1	1	1	1	1	1	1
Limerick Generating Station, Unit 2	2	2	1	1	1	1	1	1	1	1	2	2
Millstone Power Station, Unit 2	3	1	1	2	1	2	1	1	1	1	2	2
Millstone Power Station, Unit 3	1	1	1	1	1	1	1	1	1	1	1	1
Nine Mile Point Nuclear Station, Unit 1	2	2	2	1	1	1	2	1	1	1	1	2
Nine Mile Point Nuclear Station, Unit 2	2	2	2	1	1	1	1	2	2	2	1	1
Oyster Creek Nuclear Generating Station, Unit 1	2	2	1	2	3	2	2	1	2	2	1	1
Peach Bottom Atomic Power Station, Unit 2	2	2	2	2	2	1	1	1	1	1	1	1

Key: 1. Action matrix column 1: Licensee response 2. Action matrix column 2: Regulatory response 3. Action matrix column 3: Degraded cornerstone 4. Action matrix column 4: Multiple/repetitive degraded cornerstone 5. Action matrix column 5: Unacceptable performance 6. Removed from action matrix: IMC 0350 process												
	Highest level of oversight (action matrixcolumn assigned) during at least some portion of the year											
Reactor	2001	2002	2003	2004	2005	2006	2007	2008	2009	2010	2011	2012
Peach Bottom Atomic Power Station, Unit 3	2	2	2	1	1	1	1	1	1	1	1	1
Pilgrim Nuclear Power Station	1	1	1	1	1	1	1	1	1	1	2	2
R.E. Ginna Nuclear Power Plant	1	2	2	1	1	1	3	1	3	3	2	1
Salem Nuclear Generating Station, Unit 1	1	2	2	2	1	1	3	1	1	1	1	3
Salem Nuclear Generating Station, Unit 2	1	1	1	1	1	1	1	1	1	1	1	3
Seabrook Station, Unit 1	2	1	2	1	1	2	1	1	2	2	1	2
Susquehanna Steam Electric Station, Unit 1	2	2	1	1	1	1	1	1	1	2	3	3
Susquehanna Steam Electric Station, Unit 2	2	2	1	1	1	1	1	1	1	1	1	1
Three Mile Island Nuclear Station, Unit 1	2	1	1	1	2	2	1	1	1	1	1	1
Vermont Yankee Nuclear Power Plant, Unit 1	1	3	1	2	2	2	2	1	1	1	1	1
NRC Region II												
Browns Ferry Nuclear Plant, Unit 1[a]	a	a	a	a	a	a	3	2	3	4	4	4
Browns Ferry Nuclear Plant, Unit 2	1	1	1	1	1	1	2	1	3	3	1	2
Browns Ferry Nuclear Plant, Unit 3	1	1	1	1	1	1	1	1	3	3	1	2
Brunswick Steam Electric Plant, Unit 1	1	1	1	1	1	2	3	1	2	2	2	2
Brunswick Steam Electric Plant, Unit 2	1	1	1	2	2	2	2	1	2	2	2	2
Catawba Nuclear Station, Unit 1	1	1	1	1	1	1	1	1	1	1	1	2
Catawba Nuclear Station, Unit 2	1	1	1	1	1	1	1	1	1	1	1	1
Crystal River Nuclear Generating Plant, Unit 3	2	1	1	1	2	2	1	1	1	1	2	2
Edwin I. Hatch Nuclear Plant, Unit 1	1	1	1	1	2	2	1	1	2	1	1	1

Table 21. (Continued)

Key: 1. Action matrix column 1: Licensee response 2. Action matrix column 2: Regulatory response 3. Action matrix column 3: Degraded cornerstone 4. Action matrix column 4: Multiple/repetitive degraded cornerstone 5. Action matrix column 5: Unacceptable performance 6. Removed from action matrix: IMC 0350 process												
	Highest level of oversight (action matrixcolumn assigned) during at least some portion of the year											
Reactor	2001	2002	2003	2004	2005	2006	2007	2008	2009	2010	2011	2012
Edwin I. Hatch Nuclear Plant, Unit 2	1	1	1	1	2	2	2	2	3	3	1	1
H. B. Robinson Steam Electric Plant, Unit 2	1	1	1	2	1	1	1	1	1	3	3	1
Joseph M. Farley Nuclear Plant, Unit 1	2	1	1	1	1	2	3	3	2	2	1	2
Joseph M. Farley Nuclear Plant, Unit 2	2	1	1	1	1	2	3	3	2	2	1	2
McGuire Nuclear Station, Unit 1	1	1	1	1	1	1	1	2	2	2	1	1
McGuire Nuclear Station, Unit 2	1	1	1	1	1	1	1	2	2	2	1	1
North Anna Power Station, Unit 1	1	1	1	1	1	1	1	1	1	1	2	2
North Anna Power Station, Unit 2	2	1	1	1	1	1	1	1	1	2	2	2
Oconee Nuclear Station, Unit 1	3	4	2	3	2	3	3	2	2	3	1	1
Oconee Nuclear Station, Unit 2	2	1	2	3	2	3	3	1	1	3	1	1
Oconee Nuclear Station, Unit 3	2	2	2	3	2	3	3	1	1	3	1	1
Sequoyah Nuclear Plant, Unit 1	2	2	1	2	2	1	1	1	1	1	2	2
Sequoyah Nuclear Plant, Unit 2	2	2	1	1	1	1	1	1	1	1	1	1
Shearon Harris Nuclear Power Plant, Unit 1	2	3	2	2	1	1	1	1	1	1	1	2
St. Lucie Plant, Unit 1	1	1	1	1	1	1	1	1	3	3	1	3
St. Lucie Plant, Unit 2	1	1	2	2	1	1	1	1	1	1	1	1
Surry Nuclear Power Station, Unit 1	2	2	2	3	1	2	2	1	1	1	1	1
Surry Nuclear Power Station, Unit 2	2	2	2	2	1	2	2	1	1	1	1	1
Turkey Point Nuclear Generating, Unit 3	1	1	1	1	2	2	2	1	2	2	2	2
Turkey Point Nuclear Generating, Unit 4	1	1	1	1	2	2	1	1	1	2	2	2

Key:
1. Action matrix column 1: Licensee response
2. Action matrix column 2: Regulatory response
3. Action matrix column 3: Degraded cornerstone
4. Action matrix column 4: Multiple/repetitive degraded cornerstone
5. Action matrix column 5: Unacceptable performance
6. Removed from action matrix: IMC 0350 process

	Highest level of oversight (action matrixcolumn assigned) during at least some portion of the year											
Reactor	2001	2002	2003	2004	2005	2006	2007	2008	2009	2010	2011	2012
Virgil C. Summer Nuclear Station, Unit 1	2	1	1	1	1	2	2	1	1	1	1	1
Vogtle Electric Generating Plant, Unit 1	1	1	1	1	1	2	2	2	1	1	1	1
Vogtle Electric Generating Plant, Unit 2	1	1	1	1	1	2	2	2	1	1	1	1
Watts Bar Nuclear Plant, Unit 1	1	1	1	2	2	1	1	1	1	1	1	2
NRC Region III												
Braidwood Station, Unit 1	2	3	2	2	1	2	2	1	2	2	1	1
Braidwood Station, Unit 2	1	1	1	1	1	2	2	1	1	1	1	2
Byron Station, Unit 1	1	1	1	1	1	1	1	2	2	1	1	1
Byron Station, Unit 2	1	1	1	1	1	2	2	2	2	1	2	1
Clinton Power Station, Unit 1	2	2	1	1	1	1	2	1	1	1	1	1
Davis-Besse Nuclear Power Station, Unit 1[b]	1	6	6	6	6	1	1	1	2	2	1	2
Donald C. Cook Nuclear Power Plant, Unit 1	6	2	2	2	1	1	3	1	1	1	1	1
Donald C. Cook Nuclear Power Plant, Unit 2	6	3	3	3	2	1	3	1	1	1	1	1
Dresden Nuclear Power Station, Unit 2	1	1	1	2	1	2	2	1	1	1	1	1
Dresden Nuclear Power Station, Unit 3	1	2	2	2	1	1	1	1	2	2	1	1
Duane Arnold Energy Center	1	1	1	1	1	1	2	1	2	2	1	1
Fermi, Unit 2	2	2	2	2	2	2	1	1	1	1	1	2
Kewaunee Power Station	3	2	2	2	3	3	3	2	2	1	1	1
LaSalle County Station, Unit 1	1	1	1	1	1	1	1	1	1	1	1	1
LaSalle County Station, Unit 2	2	2	1	1	1	1	1	1	1	1	1	1

Table 21. (Continued)

Key:
1. Action matrix column 1: Licensee response
2. Action matrix column 2: Regulatory response
3. Action matrix column 3: Degraded cornerstone
4. Action matrix column 4: Multiple/repetitive degraded cornerstone
5. Action matrix column 5: Unacceptable performance
6. Removed from action matrix: IMC 0350 process

	Highest level of oversight (action matrixcolumn assigned) during at least some portion of the year											
Reactor	2001	2002	2003	2004	2005	2006	2007	2008	2009	2010	2011	2012
Monticello Nuclear Generating Plant, Unit 1	1	1	1	1	1	1	1	1	1	1	1	1
Palisades Nuclear Plant	2	2	2	2	1	1	1	2	2	2	3	3
Perry Nuclear Power Plant, Unit 1	1	2	3	4	4	4	4	2	1	1	3	3
Point Beach Nuclear Plant, Unit 1	2	2	4	4	4	4	1	1	1	1	1	2
Point Beach Nuclear Plant, Unit 2	2	2	4	4	4	4	1	1	1	1	1	2
Prairie Island Nuclear Generating Plant, Unit 1	2	2	1	1	1	1	1	2	2	1	2	2
Prairie Island Nuclear Generating Plant, Unit 2	2	2	1	1	1	1	1	1	2	2	1	2
Quad Cities Nuclear Power Station, Unit 1	2	1	1	1	1	2	2	1	1	1	1	1
Quad Cities Nuclear Power Station, Unit 2	2	1	1	1	1	1	1	1	1	1	1	1
NRC Region IV												
Arkansas Nuclear One, Unit 1	1	1	1	2	1	1	1	1	2	1	1	1
Arkansas Nuclear One, Unit 2	1	1	1	1	1	1	1	1	1	1	1	1
Callaway Plant	3	2	2	2	1	2	1	1	1	2	1	2
Columbia Generating Station, Unit 2	3	3	1	1	2	2	1	1	2	2	1	3
Comanche Peak Steam Electric Station, Unit 1	1	2	1	2	1	1	1	2	1	1	1	1
Comanche Peak Steam Electric Station, Unit 2	1	2	1	1	1	1	1	1	1	1	1	1

Key:
1. Action matrix column 1: Licensee response
2. Action matrix column 2: Regulatory response
3. Action matrix column 3: Degraded cornerstone
4. Action matrix column 4: Multiple/repetitive degraded cornerstone
5. Action matrix column 5: Unacceptable performance
6. Removed from action matrix: IMC 0350 process

	Highest level of oversight (action matrixcolumn assigned) during at least some portion of the year											
Reactor	2001	2002	2003	2004	2005	2006	2007	2008	2009	2010	2011	2012
Cooper Nuclear Station	3	4	4	4	1	1	2	3	2	1	2	1
Diablo Canyon Nuclear Power Plant, Unit 1	1	1	1	1	2	1	1	1	1	1	1	1
Diablo Canyon Nuclear Power Plant, Unit 2	1	1	1	1	2	1	1	1	1	1	1	1
Fort Calhoun Station, Unit 1[c]	1	2	2	2	2	1	3	3	1	3	6	6
Grand Gulf Nuclear Station, Unit 1	1	1	1	1	1	1	1	2	1	1	1	1
Palo Verde Nuclear Generating Station, Unit 1	1	1	1	1	3	3	3	3	3	1	2	2
Palo Verde Nuclear Generating Station, Unit 2	1	1	1	1	3	3	3	3	3	1	2	2
Palo Verde Nuclear Generating Station, Unit 3	1	1	1	1	3	4	4	4	4	1	2	2
River Bend Station, Unit 1	1	2	1	1	2	1	1	2	1	1	1	2
San Onofre Nuclear Generating Station, Unit 2	1	1	2	2	2	1	1	2	2	2	1	2
San Onofre Nuclear Generating Station, Unit 3	1	1	1	1	1	1	1	1	1	1	1	2
South Texas Project, Unit 1	1	1	1	1	1	1	1	1	1	1	1	1
South Texas Project, Unit 2	1	2	2	1	1	2	1	1	1	1	1	1
Waterford Steam Electric Station, Unit 3	1	1	1	2	1	1	1	1	2	2	2	2
Wolf Creek Generating Station, Unit 1	1	1	1	1	1	1	1	1	2	3	3	3

Source: GAO analysis of NRC data.

Notes: In February 2013, the owner of the Crystal River Nuclear Plant in Florida permanently shut down that site's reactor; in May 2013, the owner of the Kewaunee Power Station in Wisconsin permanently shut down that site's reactor; and in June 2013, the owner of the San Onofre Nuclear Generating Station in California permanently shut down that site's two reactors. These actions reduced the number of operating commercial nuclear power reactors in the United States from 104 to 100, and they reduced the number of operating nuclear power plant sites from 65 to 62.

NRC periodically removes plants from oversight under the action matrix—due to significant performance or operational concerns regarding plants that are shut down—through a prescribed process that it refers to as the IMC 0350 process, which is named after the NRC *Inspection Manual* chapter providing guidance for this level of oversight. Upon implementation of this process, the appropriate regional administrator establishes an oversight panel—typically comprised of officials from both headquarters and regional offices, including senior management—that determines the oversight activities necessary to authorize the restart of the plant's reactors. See NRC *Inspection Manual* chapter 0350, "Oversight of Reactor Facilities in a Shutdown Condition due to Significant Performance and/or Operational Concerns."

[a]One of the Browns Ferry reactors (Unit 1) was shut down from March 1985 until May 2007 to address performance and management issues.

[b]NRC removed the Davis-Besse Nuclear Power Station from oversight under the action matrix and placed it under the IMC 0350 process from 2002 through 2005 due to an incident with the reactor's vessel head that occurred in 2002.

[c]NRC removed the Fort Calhoun Station from oversight under the action matrix and placed it under the IMC 0350 process in 2011 due to performance issues, as well as issues associated with the Missouri River flooding that affected the plant that year. As of July 2013, this plant has not been reinstated for oversight under the action matrix.

APPENDIX VIII: SUBSTANTIVE CROSS-CUTTING ISSUES IDENTIFIED BY NRC AT OPERATING NUCLEAR POWER REACTORS IN THE UNITED STATES

This appendix provides performance information about the 104 commercial nuclear power reactors at 65 power plants for which we analyzed NRC data and that were operating in the United States through the end of calendar year 2012. Table 22 lists the substantive cross-cutting issues assigned to each reactor from 2001 through 2012.

Table 22. Substantive Cross-cutting Issues Identified by NRC at Operating Commercial Nuclear Power Reactors for One or More Quarters of Each Calendar Year, 2001-2012

Key:
HU: human performance
PIR: problem identification and resolution
SC: safety-conscious work environment

Reactor	**2001**	**2002**	**2003**	**2004**	**2005**	**2006**	**2007**	**2008**	**2009**	**2010**	**2011**	**2012**
NRC Region I												
Beaver Valley Power Station, Unit 1												
Beaver Valley Power Station, Unit 2												
Calvert Cliffs Nuclear Power Plant, Unit 1		PIR										
Calvert Cliffs Nuclear Power Plant, Unit 2		PIR										
Hope Creek Generating Station, Unit 1			PIR	PIR SC	PIR SC	SC						
Indian Point Nuclear Generating, Unit 2		PIR HU	PIR HU	PIR	PIR		HU	PIR HU	HU			
Indian Point Nuclear Generating, Unit 3								HU	HU			
James A. FitzPatrick Nuclear Power Plant	PIR											
Limerick Generating Station, Unit 1												
Limerick Generating Station, Unit 2												
Millstone Power Station, Unit 2	PIR		PIR									
Millstone Power Station, Unit 3												
Nine Mile Point Nuclear Station, Unit 1			PIR	PIR								
Nine Mile Point Nuclear Station, Unit 2			PIR	PIR								
Oyster Creek Nuclear Generating Station, Unit 1	PIR		HU		PIR	HU	HU					
Peach Bottom Atomic Power Station, Unit 2				PIR								

Table 22. (Continued)

Key:
HU: human performance
PIR: problem identification and resolution
SC: safety-conscious work environment

Reactor	2001	2002	2003	2004	2005	2006	2007	2008	2009	2010	2011	2012
Peach Bottom Atomic Power Station, Unit 3				PIR								
Pilgrim Nuclear Power Station												
R.E. Ginna Nuclear Power Plant												
Salem Nuclear Generating Station, Unit 1			PIR	PIR SC	PIR SC	SC	HU	HU	HU			
Salem Nuclear Generating Station, Unit 2			PIR	PIR SC	PIR SC	SC	HU	HU	HU			
Seabrook Station, Unit 1	PIR	PIR										
Susquehanna Steam Electric Station, Unit 1			HU	PIR							PIR	HU PIR
Susquehanna Steam Electric Station, Unit 2			HU	PIR							PIR	HU PIR
Three Mile Island Nuclear Station, Unit 1		HU		PIR								
Vermont Yankee Nuclear Power Plant, Unit 1												
NRC Region II												
Browns Ferry Nuclear Plant, Unit 1a	a	a	a	a	a	a		PIR	PIR	PIR	PIR	PIR
Browns Ferry Nuclear Plant, Unit 2								PIR	PIR	PIR	PIR	PIR
Browns Ferry Nuclear Plant, Unit 3								PIR	PIR	PIR	PIR	PIR
Brunswick Steam Electric Plant, Unit 1												
Brunswick Steam Electric Plant, Unit 2												
Catawba Nuclear Station, Unit 1												
Catawba Nuclear Station, Unit 2												

Key:
HU: human performance
PIR: problem identification and resolution
SC: safety-conscious work environment

Reactor	2001	2002	2003	2004	2005	2006	2007	2008	2009	2010	2011	2012
Crystal River Nuclear Generating Plant, Unit 3												
Edwin I. Hatch Nuclear Plant, Unit 1												
Edwin I. Hatch Nuclear Plant, Unit 2												
H. B. Robinson Steam Electric Plant, Unit 2											HU	HU
Joseph M. Farley Nuclear Plant, Unit 1										HU		
Joseph M. Farley Nuclear Plant, Unit 2										HU		
McGuire Nuclear Station, Unit 1												
McGuire Nuclear Station, Unit 2												
North Anna Power Station, Unit 1								HU				
North Anna Power Station, Unit 2								HU				
Oconee Nuclear Station, Unit 1											HU	
Oconee Nuclear Station, Unit 2											HU	
Oconee Nuclear Station, Unit 3											HU	
Sequoyah Nuclear Plant, Unit 1												
Sequoyah Nuclear Plant, Unit 2												
Shearon Harris Nuclear Power Plant, Unit 1												
St. Lucie Plant, Unit 1												
St. Lucie Plant, Unit 2												
Surry Nuclear Power Station, Unit 1												
Surry Nuclear Power Station, Unit 2												
Turkey Point Nuclear Generating, Unit 3			PIR	PIR		PIR	PIR	PIR				
Turkey Point Nuclear Generating, Unit 4			PIR	PIR		PIR	PIR	PIR				
Virgil C. Summer Nuclear Station, Unit 1												

Table 22. (Continued)

Key:
HU: human performance
PIR: problem identification and resolution
SC: safety-conscious work environment

Reactor	2001	2002	2003	2004	2005	2006	2007	2008	2009	2010	2011	2012
Vogtle Electric Generating Plant, Unit 1												
Vogtle Electric Generating Plant, Unit 2												
Watts Bar Nuclear Plant, Unit 1					HU							
NRC Region III												
Braidwood Station, Unit 1									HU	HU		
Braidwood Station, Unit 2									HU	HU		
Byron Station, Unit 1					HU	HU		HU	HU			
Byron Station, Unit 2					HU	HU		HU	HU			
Clinton Power Station, Unit 1												
Davis-Besse Nuclear Power Station, Unit 1												HU
Donald C. Cook Nuclear Power Plant, Unit 1		PIR	PIR									
Donald C. Cook Nuclear Power Plant, Unit 2		PIR	PIR									
Dresden Nuclear Power Station, Unit 2				HU	HU					HU		
Dresden Nuclear Power Station, Unit 3				HU	HU					HU		
Duane Arnold Energy Center					HU	HU						
Fermi, Unit 2					HU	PIR		HU				
Kewaunee Power Station	PIR				PIR	PIR HU	PIR HU	PIR HU				
LaSalle County Station, Unit 1		HU		HU	HU							
LaSalle County Station, Unit 2		HU		HU	HU							

Key:
HU: human performance
PIR: problem identification and resolution
SC: safety-conscious work environment

Reactor	**2001**	**2002**	**2003**	**2004**	**2005**	**2006**	**2007**	**2008**	**2009**	**2010**	**2011**	**2012**
Monticello Nuclear Generating Plant, Unit 1									HU			
Palisades Nuclear Plant	PIR HU	PIR HU	PIR				PIR	HU	HU			HU
Perry Nuclear Power Plant, Unit 1				PIR	PIR HU	PIR HU		HU	PIR HU	PIR HU	HU	HU
Point Beach Nuclear Plant, Unit 1		PIR	PIR HU	PIR HU	PIR HU		HU	PIR HU	PIR HU			
Point Beach Nuclear Plant, Unit 2		PIR	PIR HU	PIR HU	PIR HU		HU	PIR HU	PIR HU			
Prairie Island Nuclear Generating Plant, Unit 1									HU	HU		HU
Prairie Island Nuclear Generating Plant, Unit 2									HU	HU		HU
Quad Cities Nuclear Power Station, Unit 1		HU	HU				HU	HU				
Quad Cities Nuclear Power Station, Unit 2		HU	HU				HU	HU				
NRC Region IV												
Arkansas Nuclear One, Unit 1				PIR	PIR			HU				HU
Arkansas Nuclear One, Unit 2				PIR	PIR			HU				HU
Callaway Plant		PIR	PIR	PIR	HU	HU	PIR					
Columbia Generating Station, Unit 2		HU	HU	PIR	HU	PIR						HU
Comanche Peak Steam Electric Station, Unit 1												
Comanche Peak Steam Electric Station, Unit 2												

Table 22. (Continued)

Key:
HU: human performance
PIR: problem identification and resolution
SC: safety-conscious work environment

Reactor	2001	2002	2003	2004	2005	2006	2007	2008	2009	2010	2011	2012
Cooper Nuclear Station	PIR	PIR HU	PIR HU	PIR HU	PIR HU	PIR	HU	HU		HU	HU	PIR HU
Diablo Canyon Nuclear Power Plant, Unit 1			HU	PIR HU	PIR					PIR	PIR	
Diablo Canyon Nuclear Power Plant, Unit 2			HU	PIR HU	PIR					PIR	PIR	
Fort Calhoun Station, Unit 1							HU	HU				
Grand Gulf Nuclear Station, Unit 1								HU				
Palo Verde Nuclear Generating Station, Unit 1					PIR HU	PIR HU	PIR HU	PIR HU	PIR HU			
Palo Verde Nuclear Generating Station, Unit 2					PIR HU	PIR HU	PIR HU	PIR HU	PIR HU			
Palo Verde Nuclear Generating Station, Unit 3					PIR HU	PIR HU	PIR HU	PIR HU	PIR HU			
River Bend Station, Unit 1						PIR	PIR HU					
San Onofre Nuclear Generating Station, Unit 2								PIR HU	PIR HU	PIR HU	HU	
San Onofre Nuclear Generating Station, Unit 3								PIR HU	PIR HU	PIR HU	HU	
South Texas Project, Unit 1												
South Texas Project, Unit 2												

Reactor	2001	2002	2003	2004	2005	2006	2007	2008	2009	2010	2011	2012
Waterford Steam Electric Station, Unit 3			PIR									
Wolf Creek Generating Station, Unit 1								PIR HU	PIR HU	PIR	PIR HU	PIR HU

Key:
HU: human performance
PIR: problem identification and resolution
SC: safety-conscious work environment

Source: NRC.

Note: This table includes data on the 104 commercial nuclear power reactors operating in the United States through the end of calendar year 2012. In February 2013, the owner of the Crystal River Nuclear Plant in Florida permanently shut down that site's reactor; in May 2013, the owner of the Kewaunee Power Station in Wisconsin permanently shut down that site's reactor; and in June 2013, the owner of the San Onofre Nuclear Generating Station in California permanently shut down that site's two reactors. These actions reduced the number of operating commercial nuclear power reactors in the United States from 104 to 100.

[a]One of the Browns Ferry reactors (Unit 1) was shut down from March 1985 until May 2007 to address performance and management issues.

Appendix IX: NRC Escalated Findings at Operating Nuclear Power Plants in the United States

This appendix provides performance information about the 104 commercial nuclear power reactors at 65 power plants for which we analyzed NRC data and that were operating in the United States through the end of calendar year 2012. Table 23 lists the number of escalated findings identified at each plant from 2000 through 2012.

Table 23. NRC Escalated Findings (Greater-than-Green Inspection Findings and Severity Level I, II, and III Violations) per Operating Commercial Nuclear Power Plant, Calendar Years 2000-2012

Plant[a]	2000	2001	2002	2003	2004	2005	2006	2007	2008	2009	2010	2011	2012	Total
NRC Region I														
Beaver Valley Power Station	0	0	1	1	0	0	1	0	0	0	0	0	0	3
Calvert Cliffs Nuclear Power Plant	0	1	2	0	1	0	1	0	0	1	1	0	0	7
Hope Creek Generating Station	0	0	0	0	2	0	0	0	0	0	0	0	0	2
Indian Point Nuclear Generating[b]	0	2	1	0	0	1	0	0	0	0	0	0	0	4
James A. Fitzpatrick Nuclear Power Plant	0	0	0	0	0	0	0	0	0	0	0	1	0	1
Limerick Generating Station	0	3	0	0	0	0	0	0	0	0	0	1	0	4
Millstone Power Station	1	0	0	0	0	0	0	0	0	0	0	1	0	2
Nine Mile Point Nuclear Station	0	0	0	1	0	0	0	1	0	1	0	0	0	3
Oyster Creek Nuclear Generating Station	0	0	0	0	2	1	0	0	0	0	0	0	0	3
Peach Bottom Atomic Power Station	0	2	1	1	0	0	0	0	0	0	0	0	0	4
Pilgrim Nuclear Power Station	0	0	0	0	0	1	0	0	0	0	0	1	0	2
R.E. Ginna Nuclear Power Plant	0	0	1	0	0	0	0	0	1	2	0	0	0	4

Planta	2000	2001	2002	2003	2004	2005	2006	2007	2008	2009	2010	2011	2012	Total
Salem Nuclear Generating Station	1	0	0	1	0	0	0	0	0	0	0	0	0	2
Planta	2000	2001	2002	2003	2004	2005	2006	2007	2008	2009	2010	2011	2012	Total
Seabrook Station	0	1	0	0	0	0	0	0	0	1	0	0	1	3
Susquehanna Steam Electric Station	1	1	1	0	0	0	0	0	0	1	1	0	0	5
Three Mile Island Nuclear Station	0	1	0	0	0	1	0	0	0	0	0	0	0	2
Vermont Yankee Nuclear Power Plant	0	0	0	0	2	0	1	0	0	0	0	0	0	3
Total (NRC Region I)	3	11	7	4	7	4	3	1	1	6	2	4	1	54
NRC Region II														
Browns Ferry Nuclear Plant[c]	1	1	0	0	0	0	0	0	0	2	1	0	1	6
Brunswick Steam Electric Plant	0	0	0	0	1	0	0	1	0	1	1	1	0	5
Catawba Nuclear Station	0	0	0	0	0	1	0	0	0	0	0	0	1	2
Crystal River Nuclear Generating Plant	0	0	0	0	0	1	0	0	0	0	0	1	0	2
Joseph M. Farley Nuclear Plant	0	0	0	0	0	0	0	2	1	1	0	0	0	4
Edwin I. Hatch Nuclear Plant	0	0	0	0	0	1	0	0	0	2	0	0	0	3
McGuire Nuclear Station	0	0	0	0	0	0	0	0	1	0	0	0	0	1
North Anna Power Station	0	0	0	0	0	0	0	0	0	0	0	1	0	1

Table 23. (Continued)

Plant[a]	2000	2001	2002	2003	2004	2005	2006	2007	2008	2009	2010	2011	2012	Total
Oconee Nuclear Station	0	2	3	1	1	0	2	0	1	0	3	1	0	14
H.B. Robinson Steam Electric Plant	0	0	0	0	0	0	0	0	0	0	3	0	0	3
St Lucie Plant	0	0	0	0	0	0	0	0	0	1	0	0	0	1
Sequoyah Nuclear Plant	0	1	0	0	1	0	0	0	0	0	0	0	0	2
Shearon Harris Nuclear Power Plant	1	1	1	0	0	0	0	0	0	0	0	0	1	4
Surry Nuclear Power Station	0	1	0	0	1	0	1	0	0	0	0	0	0	3
Turkey Point Nuclear Generating	0	0	0	0	0	1	1	0	0	2	0	1	0	5
Virgil C. Summer Nuclear Station	1	1	0	0	0	0	1	1	0	0	0	0	0	4
Vogtle Electric Generating Plant	0	0	0	0	0	0	1	0	0	0	0	0	0	1
Watts Bar Nuclear Plant	0	1	0	0	1	1	0	0	0	0	0	0	0	3
Total (NRC Region II)	3	8	4	1	5	5	6	4	3	9	8	5	3	64
NRC Region III														
Braidwood Station	0	0	1	0	0	0	1	0	0	1	0	0	0	3
Byron Station	0	0	0	0	0	1	0	0	1	0	0	1	0	3
Clinton Power Station	0	1	0	1	0	0	1	0	0	0	0	0	0	3
Davis-Besse Nuclear Power Station	1	0	7	3	1	0	0	0	0	1	1	0	0	14

Plant[a]	2000	2001	2002	2003	2004	2005	2006	2007	2008	2009	2010	2011	2012	Total
Donald C. Cook Nuclear Power Plant	0	0	2	0	2	6	1	0	0	0	0	0	0	11
Dresden Nuclear Power Station	0	2	0	1	0	0	0	0	0	1	0	0	0	4
Duane Arnold Energy Center	0	0	0	0	0	0	3	0	0	1	0	0	0	4
Fermi	0	1	0	0	0	0	0	0	0	0	0	0	0	1
Kewaunee Power Station	1	1	1	0	1	2	0	1	1	0	0	0	1	9
Lasalle County Station	0	0	0	0	0	1	1	0	0	0	0	0	0	2
Monticello Nuclear Generating Plant	0	0	0	0	0	0	0	0	0	1	0	0	0	1
Palisades Nuclear Plant	0	2	0	1	0	0	0	0	1	1	0	3	0	8
Perry Nuclear Power Plant	0	1	1	3	1	0	0	0	0	0	0	1	1	8
Point Beach Nuclear Plant	0	1	3	3	0	2	1	0	0	1	0	0	2	13
Prairie Island Nuclear Generating Plant	0	1	0	0	0	0	0	0	1	3	0	1	1	7
Quad Cities Nuclear Power Station	1	0	0	0	0	0	1	0	0	0	0	0	0	2
Total (NRC Region III)	3	10	15	12	5	12	9	1	4	10	1	6	5	93
NRC Region IV														
Arkansas Nuclear One	0	1	0	0	0	0	0	0	0	0	0	0	0	1

Table 23. (Continued)

Plant[a]	2000	2001	2002	2003	2004	2005	2006	2007	2008	2009	2010	2011	2012	Total
Callaway Plant	3	2	1	1	0	0	0	0	0	0	0	0	0	7
Columbia Generating Station	0	1	1	0	0	0	0	0	0	0	0	0	2	4
Comanche Peak Steam Electric Station	0	1	1	0	0	0	0	0	0	0	0	0	0	2
Cooper Nuclear Station	1	1	3	0	1	0	0	1	2	0	0	2	0	11
Diablo Canyon Nuclear Power Plant	0	0	0	0	0	0	0	0	0	0	0	0	0	0
Fort Calhoun Station	0	0	1	0	0	1	1	2	0	0	1	1	1	8
Grand Gulf Nuclear Station	0	0	0	0	0	0	0	0	0	0	0	0	0	0
Palo Verde Nuclear Generating Station	0	0	0	0	2	1	2	0	0	0	0	0	0	5
River Bend Station	0	0	2	0	0	0	0	0	0	0	0	0	0	2
San Onofre Nuclear Generating Station	0	0	0	0	0	0	0	0	1	0	0	1	0	2
South Texas Project	0	0	0	0	0	0	0	0	0	0	0	0	0	0
Waterford Steam Electric Station	0	0	0	0	1	0	0	0	0	1	0	1	0	3
Wolf Creek Generating Station	0	0	0	0	0	0	0	0	0	0	0	0	1	1
Total (NRC Region IV)	4	6	9	1	4	2	3	3	3	1	1	5	4	46
Total (all regions)	13	35	35	18	21	23	21	9	11	26	12	20	13	257

Source: GAO analysis of NRC data.

Notes: This table includes data on the 104 commercial nuclear power reactors operating in the United States through the end of calendar year 2012. In February 2013, the owner of the Crystal River Nuclear Plant in Florida permanently shut down that site's reactor; in May 2013, the owner of the Kewaunee Power Station in Wisconsin permanently shut down that site's reactor; and in June 2013, the owner of the San Onofre Nuclear Generating Station in California permanently shut down that site's two reactors. These actions reduced the number of operating commercial nuclear power reactors in the United States from 104 to 100.

Greater-than-green findings refer to white, yellow, and red findings. White findings are denoted as low to moderate safety significance; they indicate that licensee performance is acceptable, but outside the nominal risk range. Cornerstone objectives are met with minimal reduction in safety margin. Yellow findings are denoted as substantial safety significance; they indicate a decline in licensee performance that is still acceptable with cornerstone objectives met, but with significant reduction in safety margin. Red findings are denoted as high safety significance; they indicate a decline in licensee performance that is associated with an unacceptable loss of safety margin. Sufficient safety margin still exists to prevent undue risk to public health and safety.

Severity Level I violations are violations that resulted in or could have resulted in serious safety consequences (e.g., violations that created the substantial potential for safety consequences or violations that involved systems failing when actually called on to prevent or mitigate a serious safety event). Severity Level II violations are violations that resulted in or could have resulted in significant safety consequences (e.g., violations that created the potential for substantial safety consequences or violations that involved systems not being capable, for an extended period, of preventing or mitigating a serious safety event). Severity Level III violations are violations that resulted in or could have resulted in moderate safety consequences (e.g., violations that created a potential for moderate safety consequences or violations that involved systems not being capable, for a relatively short period, of preventing or mitigating a serious safety event).

[a] NRC reports these data by nuclear power plant as opposed to by individual reactor. Oftentimes, there are 2 or 3 reactors located at each plant. Therefore, data for all 104 reactors are included here, but at the plant level.

[b] Although NRC tracks findings at Indian Point Units 2 and 3 separately, they are physically located at the same plant.

[c] One of the Browns Ferry reactors (Unit 1) was shut down from March 1985 until May 2007 to address performance and management issues.

APPENDIX X: NRC FUKUSHIMA NEAR-TERM TASK FORCE RECOMMENDATIONS

NRC convened the Fukushima Near-Term Task Force to review its processes and regulations and determine whether its oversight could be informed by lessons learned from the March 2011 disaster at the Fukushima Daiichi Nuclear Power Plant in Japan. The task force concluded that a sequence of events like the Fukushima accident is unlikely to occur in the United States and that the continued operation of the nation's commercial nuclear reactors does not pose an imminent risk to public health and safety. To improve NRC's oversight, the task force recommended:

1. Establishing a logical, systematic, and coherent regulatory framework for adequate protection that appropriately balances defense-in-depth and risk considerations.
2. NRC require licensees to reevaluate and upgrade as necessary the design-basis seismic and flooding protection of structures, systems, and components for each operating reactor.
3. As part of the longer term review, that NRC evaluate potential enhancements to the capability to prevent or mitigate seismically induced fires and floods.
4. NRC strengthen station blackout mitigation capability at all operating and new reactors for design-basis and beyond-design-basis external events.
5. Requiring reliable hardened vent designs in boiling water reactor facilities with Mark I and Mark II containments.
6. As part of the longer term review, that NRC identify insights about hydrogen control and mitigation inside containment or in other buildings as additional information is revealed through further study of the Fukushima Dai-ichi accident.
7. Enhancing spent fuel pool makeup capability and instrumentation for the spent fuel pool.
8. Strengthening and integrating onsite emergency response capabilities such as emergency operating procedures, severe accident management guidelines, and extensive damage mitigation guidelines.
9. NRC require that facility emergency plans address prolonged station blackout and multiunit events.

10. As part of the longer term review, that NRC pursue additional emergency preparedness topics related to multiunit events and prolonged station blackout.
11. As part of the longer term review, that NRC should pursue emergency preparedness topics related to decisionmaking, radiation monitoring, and public education.
12. NRC strengthen regulatory oversight of licensee safety performance (i.e., the Reactor Oversight Process) by focusing more attention on defense-in-depth requirements consistent with the recommended defense-in-depth framework.

End Notes

[1] Low-probability, high-consequence events are accidents that are unexpected, with few similar historical events; however, when they do happen, the impact is extreme. Such events include natural disasters, terrorism, nuclear power plant accidents, and chemical plant explosions.

[2] In this report, when we use the term nuclear power plant, we are referring to an entire site that uses one or more nuclear power reactors to generate electricity.

[3] NRC, *Recommendations for Enhancing Reactor Safety in the 21st Century: The Near-Term Task Force Review of Insights from the Fukushima Dai-ichi Accident* (July 12, 2011).

[4] NRC, Office of the Inspector General, *Evaluation Report: Inspector General's Assessment of the Most Serious Management and Performance Challenges Facing NRC*, OIG-13-A-01 (Oct. 1, 2012); *Evaluation Report: Inspector General's Assessment of the Most Serious Management and Performance Challenges Facing NRC*, OIG-12-A-01 (Oct. 3, 2011); and *Evaluation Report: Inspector General's Assessment of the Most Serious Management and Performance Challenges Facing NRC*, OIG-11-A-01 (Oct. 1, 2010).

[5] NRC issues licenses for commercial nuclear power reactors to operate for up to 40 years and allows these licenses to be renewed for up to an additional 20 years. There is no set limit on the number of times a reactor's operating license can be renewed.

[6] Most of the nation's operating commercial nuclear power reactors received their initial operating licenses in the 1970s and 1980s. By the end of 2013, 62 reactors will have held an operating license for at least 30 years, and of these reactors, 20 will have held an operating license for 40 years or more. As a result, many reactors are reaching or have reached the end of their initial 40-year operating period. As of May 2013, NRC had renewed 73 reactor licenses and was reviewing license renewal applications for 14 reactors. However, in May 2013, the owner of the Kewaunee Power Station in Wisconsin permanently shut down that site's reactor, and in June 2013, the owner of the San Onofre Nuclear Generating Station in California permanently shut down that site's two reactors. The Kewaunee reactor was one of the 73 reactors with an operating license renewed by NRC, but the San Onofre reactors were not. As a result, as of July 2013, there are 72 commercial nuclear power reactors with renewed operating licenses operating in the United States.

[7] GAO, *Nuclear Reactor License Renewal: NRC Generally Follows Documented Procedures, but Its Revisions to Environmental Review Guidance Have Not Been Timely*, GAO-13-493 (Washington, D.C.: May 30, 2013).

[8] GAO, *Nuclear Regulatory Commission: Oversight of Nuclear Power Plant Safety Has Improved, but Refinements Are Needed*, GAO-06-1029 (Washington, D.C.: Sept. 27, 2006).

[9] NRC, Office of the Inspector General, *Audit of NRC's Enforcement Program*, OIG-08-A17 (Sept. 26, 2008).

[10] Because this was a nonprobability sample, the information we gathered from these visits to these plants is not generalizable to all 65 U.S. nuclear power plant sites but provides important illustrative information for understanding the oversight of reactors at the selected plants.

[11] In this report, we use "operating" to describe those commercial nuclear power reactors and plants currently licensed by NRC to operate in the United States, regardless of whether they may be temporarily shut down (outages) for refueling, maintenance, inspection, emergency, or other nonpermanent purposes. Licensees must submit formal notification to NRC to permanently shut down a reactor. In February 2013, the owner of the Crystal River Nuclear Plant in Florida permanently shut down that plant's reactor; in May 2013, the owner of the Kewaunee Power Station in Wisconsin permanently shut down that plant's reactor; and in June 2013, the owner of the San Onofre Nuclear Generating Station in California permanently shut down that plant's two reactors. These actions reduced the number of operating commercial nuclear reactors in the United States from 104 to 100, and they reduced the number of operating nuclear power plants from 65 to 62.

[12] According to NRC documents, licensees voluntarily submit performance indicator data electronically to NRC on a per-reactor basis under a program of the Nuclear Energy Institute (the commercial nuclear trade organization). The Nuclear Energy Institute develops—with input from NRC staff—and publishes the guidance for licensees to collect and report performance indicator data. According to NRC officials, NRC formally endorses the Nuclear Energy Institute's proposals.

[13] These inspections are focused on activities and systems that, if they fail, have the potential to trigger—or increase the consequences of—an accident, along with activities and systems that could mitigate the effects of such an accident.

[14] "Minor issues" are defined by NRC as those that have little actual safety consequences, little or no potential to impact safety, little impact on the regulatory process, and no willfulness. For example, if a licensee missed providing an hourly update to a state agency during an event that resulted in no actual safety consequences, it would be considered minor if it did not detract significantly from the state agency's ability to function during the emergency.

[15] The measures used for the characterization of risk are risk of damage to the reactor core and risk of release of radiation. In some situations, risk calculations cannot be made using these measures, such as in the case of measuring the risk for emergency preparedness inspection findings. In these cases, thresholds were determined by panels of experts on the basis of operating experience and a determination of the appropriate response.

[16] NRC assigns substantive cross-cutting issues on a per-plant—rather than a perreactor—basis.

[17] In some cases, NRC may determine that a finding poses a safety risk even if it does not violate a regulatory or other requirement. In other cases, a licensee's performance may violate a regulatory or other requirement, but NRC does not consider the issue to be a finding and therefore does not assign it a risk significance level (i.e., green, white, yellow, red) either because NRC has determined that the issue is a minor performance deficiency (i.e., less

than very low safety significance), is not a performance deficiency (i.e., is not reasonably within the licensee's ability to foresee and correct), or the safety risk cannot be quantified.

[18] Orders can be issued to modify, suspend, or revoke a license; to cease and desist from a given practice or activity; or take such other action as may be proper.

[19] NRC periodically removes plants from oversight under the action matrix—due to significant performance or operational concerns regarding plants that are shut down— through a prescribed process that it refers to as the IMC 0350 process, which is named after the NRC *Inspection Manual* chapter providing guidance for this level of oversight. Upon implementation of this process, the appropriate regional administrator establishes an oversight panel—typically comprised of officials from both headquarters and regional offices, including senior management—that determines the oversight activities necessary to authorize the restart of the plant's reactors.

[20] http://www.nrc.gov/NRR/OVERSIGHT/ASSESS/actionmatrix_summary.html. Accessed July 15, 2013.

[21] These are violations that resulted in actual safety or security consequences, including but not limited to violations resulting in radiation exposures to the public or plant personnel above regulatory limits; any violation during an actual general emergency that prevents offsite response organizations from implementing protective actions, under their emergency plans, to protect the public health or safety, and violations resulting in substantial releases of radioactive material; violations that may impact the ability of the NRC to perform its regulatory oversight function; violations involving willfulness; violations of NRC requirements for which there are no associated Significance Determination Process performance deficiencies.

[22] All violations are subject to consideration for civil penalties, but according to NRC policy, violations assessed under the ROP—unlike those assessed under the Traditional Enforcement Process—are not typically considered for civil penalties unless they involve actual consequences.

[23] According to NRC officials, each region has its own guidance for implementing NRC's policies.

[24] Issues determined to be minor performance deficiencies do not meet the criteria for an inspection finding.

[25] According to NRC officials, NRC rotates its resident inspectors among plants every 7 years.

[26] According to NRC's enforcement policy, a "reasonable period of time" is commensurate with the severity of the violation.

[27] A reactor at the Three Mile Island Nuclear Generating Station in southeastern Pennsylvania partially melted down on March 28, 1979. This was the most serious accident in U.S. commercial nuclear power plant operating history. In response to this accident, NRC implemented several changes to increase oversight of nuclear plants. These changes include expanding NRC's resident inspector program—first authorized in 1977—to have at least two inspectors live nearby and work exclusively at each plant in the United States to provide daily surveillance of licensee adherence to NRC regulations; and integrating NRC observations, findings, and conclusions about licensee performance and management effectiveness into periodic, public reports.

[28] NRC, in implementing requirements of the Price-Anderson Act as amended, requires all licensees of nuclear power plants to have financial protection including liability insurance to ensure that funds would be available in the event of a nuclear accident at a U.S. nuclear power plant. American Nuclear Insurers—an insurance pool comprised of various insurance companies in the United States—provides liability insurance for nuclear power plants.

[29] GAO-06-1029.

[30] According to NRC's *Inspection Manual*, the SERP provides a management review of the preliminary significance characterization and basis of findings that are potentially white, yellow, red, or greater than green. When necessary, based on the results of a Regulatory Conference or written response provided by the licensee, the SERP provides the management review of the final significance characterization and the basis of findings that are white, yellow, or red. No official agency preliminary significance determination of white, yellow, red, or greater than green is to be made without a SERP review. During the SERP, panel members will discuss the merits of the finding and reach consensus on (1) the statement of deficient licensee performance on which the inspection finding is based; (2) the safety significance of the finding, including assignment of preliminary or final color; and (3) the apparent violation and the regulatory requirements that should be cited.

[31] GAO-06-1029.

[32] According to NRC's website, the Office of Nuclear Reactor Regulation is responsible for accomplishing key components of NRC's nuclear reactor safety mission. As such, the office conducts a broad range of regulatory activities in the four primary program areas of rulemaking, licensing, oversight, and incident response for commercial nuclear power reactors, and test and research reactors to protect the public health, safety, and the environment. The office works with the NRC regions and other offices to accomplish its mission and contribute to the agency mission.

[33] In our analysis, we use "nonescalated findings" to refer to combined NRC data for (1) green findings identified through the ROP (i.e., equating to very low risk significance) and (2) Severity Level IV violations identified through the Traditional Enforcement Process (i.e., violations that are less serious, but are of more than minor concern resulting in no or relatively inappreciable potential safety consequences).

[34] In our analysis, we use "escalated findings" to refer to combined NRC data for (1) greater than green (i.e., white, yellow, and red) findings identified through the ROP and (2) escalated violations (i.e., Severity Levels III, II, and I) identified through the Traditional Enforcement Process.

[35] NRC requires inspectors to document details supporting each finding or violation in an inspection report.

[36] NRC, OIG-08-A-17.

[37] GAO, *Standards for Internal Control in the Federal Government*, GAO/AIMD-00-21.3.1 (Washington, D.C.: Nov.1999).

[38] According to NRC officials, ADAMS contains some documents dating from before November 1999.

[39] According to NRC information, every reactor site has at least one spent fuel pool into which fuel is placed for storage when it is removed from the reactor. Spent nuclear fuel refers to the bundles of uranium pellets encased in metal rods that have been used to power a nuclear reactor. Nuclear fuel loses efficiency over time and becomes unable to keep a nuclear reaction going. Periodically, about one-third of the fuel assemblies in a reactor are replaced. The nuclear reaction is stopped before the spent fuel is removed. But spent fuel still produces a lot of radiation and heat that must be managed to protect workers, the environment, and the public.

[40] In March 2002, the most serious safety issue confronting the nation's commercial nuclear power industry since Three Mile Island in 1979 was identified at the Davis-Besse plant in Ohio. After NRC allowed Davis-Besse to delay shutting down to inspect its reactor vessel

for cracked tubing, the plant found that leakage from these tubes had caused extensive corrosion on the vessel head—a vital barrier preventing a radioactive release.

[41] GAO, *Nuclear Regulation: NRC Needs to More Aggressively and Comprehensively Resolve Issues Related to the Davis-Besse Nuclear Power Plant's Shutdown*, GAO-04-415 (Washington, D.C.: May 17, 2004).

[42] GAO, *National Security: Key Challenges and Solutions to Strengthen Interagency Collaboration*, GAO-10-822T (Washington, D.C.: June 9, 2010).

End Note for Appendix I

[1] We did not analyze security findings because they are outside the scope of this review, which focuses on NRC's oversight of reactor safety, not security.

In: Nuclear Regulatory Commission
Editor: Shana Mayfield
ISBN: 978-1-62948-818-9

Chapter 2

INSPECTOR GENERAL'S ASSESSMENT OF THE MOST SERIOUS MANAGEMENT AND PERFORMANCE CHALLENGES FACING NRC*

Hubert T. Bell

ABSTRACT

In accordance with the Reports Consolidation Act of 2000, I have updated what I consider to be the most serious management and performance challenges facing the U.S. Nuclear Regulatory Commission (NRC). This report conveys the updated challenge list and provides a description of each challenge area and an assessment of agency actions to address these areas.

BACKGROUND AND OBJECTIVES

On January 24, 2000, Congress enacted the *Reports Consolidation Act of 2000*, requiring Federal agencies to provide financial and performance management information in a more meaningful and useful format for

* This memorandum was released by the U.S. Nuclear Regulatory Commission, Office of the Inspector General, October 2, 2013.

Congress, the President, and the public. The act requires the Inspector General (IG) of each Federal agency to annually summarize what he or she considers to be the most serious management and performance challenges facing the agency and to assess the agency's progress in addressing those challenges.

To accomplish this assessment, I considered the overall work of the Office of the Inspector General (OIG), the OIG staff's general knowledge of agency operations, and other relevant information to develop and update the list of management and performance challenges and assess the agency's progress in addressing these challenges. In addition, my staff sought input from NRC's Chairman, Commissioners, and management to obtain their views on what challenges the agency is facing and what efforts the agency has taken or are underway to address previously identified management and performance challenges.

RESULTS

The NRC's mission is to license and regulate the Nation's civilian use of byproduct, source, and special nuclear materials to ensure adequate protection of public health and safety, promote the common defense and security, and protect the environment. Like other Federal agencies, NRC faces management and performance challenges in carrying out its mission.

Most Serious Management and Performance Challenges Facing NRC as of October 1, 2013* (as identified by Inspector General)

Challenge 1	Management of regulatory processes to meet a changing environment in the oversight of nuclear materials.
Challenge 2	Management of NRC security programs.
Challenge 3	Management of regulatory processes to meet a changing environment in the oversight of nuclear facilities.
Challenge 4	Management of regulatory processes associated with high-level radioactive waste.
Challenge 5	Management of information technology.
Challenge 6	Administration of all aspects of financial management and procurement.
Challenge 7	Management of human capital.

*The most serious management and performance challenges are not ranked in any order of importance.

Congress left the determination and threshold of what constitutes a most serious management and performance challenge to the discretion of the IGs. I have defined serious management and performance challenges as *mission critical areas or programs that have the potential for a perennial weakness or vulnerability that, without substantial management attention, would seriously impact agency operations or strategic goals*. Based on this definition, I have identified the following as the most serious management and performance challenges facing NRC as of October 1, 2013:

CHALLENGE 1. MANAGEMENT OF REGULATORY PROCESSES TO MEET A CHANGING ENVIRONMENT IN THE OVERSIGHT OF NUCLEAR MATERIALS

Overview

NRC is responsible for maintaining an established regulatory framework for the safe and secure use of nuclear materials; medical, industrial, and academic applications; and uranium recovery activities. NRC is authorized to grant licenses for the possession and use of radioactive materials and establish regulations to govern the possession and use of those materials. Agency regulations require that certain material licensees have extensive material control and accounting programs as a condition of their licenses. Other license applicants (including those requesting authorization to possess small quantities of special nuclear materials[1]) must develop and implement plans that demonstrate a commitment to accurately control and account for radioactive materials. Upon a State's request, NRC may enter into an agreement to relinquish its authority to the State to regulate certain radioactive materials (including low-level waste) and limited quantities of special nuclear material. The State must demonstrate that its regulatory program is adequate to protect public health and safety and compatible with NRC's program. The States that enter into an agreement assuming this regulatory authority from NRC are called Agreement States. Currently, there are 37 Agreement States.

Issues

- Ensure appropriate oversight of radioactive material. This includes the accurate tracking and control of byproduct material, especially those materials with the greatest potential to impact public health and safety.
- Ensure that radioactive material is adequately protected to prevent its use for malicious purposes.
- Ensure reliable accounting of special nuclear materials in the NRC and Department of Energy jointly managed Nuclear Materials Management and Safeguards System.
- Ensure the appropriate oversight of uranium recovery facilities. The Department of Energy is responsible for cleanup and remediation of these sites under an NRC general license.
- Ensure that Agreement State programs are adequate to protect public health and safety and the environment, and are compatible with NRC's program.
- Ensure the management and safe storage and disposal of low-level radioactive waste produced as a result of NRC-licensed activities.
- Improve implementation of the National

Environmental Policy Act (NEPA). Assessment

During FY 2012, OIG audited NRC's oversight of industrial radiography and found areas where NRC could improve. In FY 2013, NRC continued work on addressing OIG's recommendations to improve its oversight of radiography.

The agency is revising its inspection guidance pertaining to radiography—although the revision has been delayed several times.

During FY 2012, OIG conducted two audits concerning general licenses.

- In the first audit, OIG identified that general licensed devices (GLD) could contain dangerous radioactive sources even though persons with no radiation training or experience were allowed to operate the devices. In response to the OIG audit, during FY 2013, agency staff contacted NRC's general licensees that possessed devices with dangerous radioactive sources and encouraged those licensees to transfer their devices to specific licenses.

- In the second audit, OIG found that many general licensees are unaware of NRC's regulatory requirements.
  The Atomic Energy Act, through the Code of Federal Regulations, establishes regulatory requirements for general licensed devices. However, NRC relies on manufacturers to make general licensees aware of these requirements, thereby delegating some of its responsibilities.
  In response to the OIG audit, during FY 2013, NRC staff committed to develop a procedure to provide each new NRC general licensee with information that will clearly describe all applicable regulatory requirements.

Since FY 2003, OIG has been tracking NRC's progress regarding the reliable accounting of special nuclear materials. While the agency continues to make progress, it still needs to complete rulemaking on Part 74, Material Control and Accounting of Special Nuclear Material. The final rule and associated guidance are scheduled to be completed by November 12, 2013, and March 12, 2014, respectively.

During FY 2012, OIG audited NRC's oversight of uranium recovery facilities and found opportunities for improvement.

In response to the OIG audit, during FY 2013, NRC staff completed actions to ensure compliance with memorandums of understanding it had with the Environmental Protection Agency and developed guidance for inspecting uranium recovery sites transferred to the Department of Energy for long-term surveillance.

Low-level radioactive waste is a byproduct of the country's use and production of certain radioactive material. NRC and the Agreement States regulate low-level waste disposal through a combination of regulatory requirements, licensing, and safety oversight. Currently, there are three low-level waste disposal sites that are regulated by Agreement States. However, these low-level waste disposal sites accept waste only from certain States or accept only limited types of low-level wastes.

Low-level radioactive waste that cannot be disposed of at a commercially operated facility is stored "onsite" where it was produced. Onsite storage increases the risk of accident and subjects workers to an increased likelihood of an unplanned exposure. NRC regulates approximately 1,600 materials licensees whose State compact does not have a low-level radioactive waste disposal facility or is not affiliated in a compact.

Challenge 2. Management of NRC Security Programs

Overview

NRC must remain vigilant with regard to the security of its infrastructure and that of nuclear facilities and nuclear materials. NRC must continue to use robust, proactive measures to protect its infrastructure – the buildings, personnel, and information – from both internal and external threats. Moreover, as the nature of the threat continues to evolve, NRC faces challenges with protecting nuclear facilities and materials, the sharing of sensitive information, as well as emergency preparedness and incident response.

Issues

- Ensure that NRC maintains strong internal physical and information security programs to protect NRC assets (e.g., NRC headquarters and regional facilities, safeguards and classified information, and information systems).
- Ensure that NRC strengthens the cyber security inspection program to protect licensee-owned assets from compromise consistent with 10 CFR Part 73 requirements.
- Ensure effective nuclear power plant emergency preparedness oversight.
- Ensure that NRC maintains a strong physical security program to protect licensee-owned assets from attack.
- Ensure implementation of the November 21, 2012, Presidential Memorandum, "National Insider Threat Policy and Minimum Standards for Executive Branch Insider Threat Programs." Currently, the staff is working toward several options for presentation to the Commission that consist of the following: (1) NRC staff informing the Commission of the framework of the NRC's insider threat

program as required by Executive Order 13587, "Structural Reforms to Improve the Security of Classified Networks and the Responsible Sharing and Safeguarding of Classified Information"; (2) a policy paper/basic implementation plan that will be a vote paper for Commission consideration, which will include policy questions including the allocation of resources and implications for bargaining unit employees; and (3) once the National Industrial Security Program Operating Manual is finalized, another paper to consider whether and to what degree licensees will be covered by the insider threat program.

Ensure that NRC continually enhances its capabilities to deter and defeat external threats to its cyber infrastructure to thwart individuals and groups that deploy malicious malware and offensive cyber capabilities for the purpose of accessing NRC's domain for malevolent purposes. Additionally, spear phishing attacks have increased at NRC and present an infrastructure information assurance challenge for the agency.

Assessment

During FY 2013, OIG evaluated NRC's compliance with the Federal Information Systems Management Act of 2002 (FISMA).[2] NRC has continued to make improvements to its information technology security program and progress in implementing the recommendations from previous FISMA evaluations. Most notable is that the agency has continued to maintain authorizations to operate on its agency and contractor systems.[3]

During FY 2013, OIG was mandated to audit NRC's classified information security program's policies and procedures. OIG found a limited number of marking errors but no evidence of systemic misclassification. Yet, this audit did identify several areas where compliance with Federal Government standards could be improved. For example, OIG found that some NRC document classifiers had not received required training; trained classifiers are not issued documentation that training was received; required classification self-assessments would be enhanced by representative sampling; and NRC's policy guidance on classification is outdated. NRC is taking action on the recommendations to strengthen the classified information security program.

Additionally, OIG audited NRC's process for ensuring that NRC employees are complying with personnel reporting responsibilities for continued NRC access authorization eligibility. During this audit, OIG found that NRC had implemented a new system to track personnel security background investigations. This new system should lead to efficiencies in the work. Nevertheless, OIG found that employees are not complying with personnel reporting responsibilities for continued NRC access authorization eligibility. OIG found that NRC employees rarely self-report the occurrence of certain events or conduct that may bring into question their reliability and trustworthiness even though such reporting is a requirement for continued NRC access authorization. OIG made recommendations to strengthen the compliance process. NRC was in agreement with the recommendations and is taking corrective actions.

Challenge 3. Management of Regulatory Processes to Meet a Changing Environment in the Oversight of Nuclear Facilities

Overview

NRC faces the challenge of maintaining its core regulatory programs while adapting to changes in its regulatory environment. NRC must address a highly variable interest in licensing and constructing new nuclear power plants to meet the Nation's increasing demands for energy production. As of May 2013, NRC had received 18 Combined License applications, 10 of which NRC was actively reviewing. Moreover, the agency is reviewing two standard design certifications and, for advanced reactors, expects to receive three design certification applications and one construction permit application through 2015.

While responding to the emerging demands associated with licensing and regulating new reactors, NRC must maintain focus and effectively carry out its current regulatory responsibilities, such as inspections of the current fleet of operating nuclear reactors and fuel cycle facilities. NRC intends to increase its safety focus on licensing and oversight activities through risk-informed and performance-based regulation.

Issues

New Facilities

- Implement the new Construction Inspection Program.
  - Risk-inform Construction Inspection Program activities to ensure the safe operation of newly constructed nuclear facilities.
  - Ensure that the NRC staff has the necessary knowledge and skill to successfully implement the program.
- As the search for new energy sources continues, NRC must ensure that the process for reviewing applications for new nuclear facilities focuses on safety and effectiveness.
- As the sources of manufactured reactor components become more globalized, NRC must ensure that its regulations and oversight activities appropriately address the challenges associated with licensees procuring components from suppliers located outside the United States.

Existing Fleet

- Ensure that NRC maintains the ability to effectively review licensee applications for license renewals and power uprates submitted by industry in response to the Nation's increasing demands for energy production.
- Ensure the most important operational issues remain the agency's top priority, and consistently apply responsive regulatory and review changes across the existing fleet of reactors.
- Establish and maintain effective, stable, and predictable regulatory programs or policies for all program areas.

Cross-Cutting Issues

- Systematically identify and develop — and consistently implement — internal controls to ensure effectiveness and efficiency of agency operations and resources.
- Identify and improve weak, informal, or unstructured processes to facilitate effective, efficient, and consistent staff activities.
- Improve the clarity, consistency, and comprehensiveness of guidance for NRC staff, such as the Inspection Manual and inter/intra-office guidance.

Assessment

OIG conducted audit followup activities in a number of significant issue areas, including oversight of licensees' vendors and reporting of nuclear power plant component defects, enforcement mechanisms, and new construction inspection. OIG auditors concurred with agency actions to close out recommendations in previously audited areas, including agency efforts to:

- Develop and implement a quality assurance process that ensures the accuracy and completion of enforcement data.
- Develop a vendor inspection program planning document that clearly articulates the program purpose and establishes metrics to evaluate program performance.
- Update inspection procedures to ensure NRC inspectors are evaluating licensee component defect reporting issues on a continual basis.

However, significant audit recommendations remain open regarding the clarity of licensee component defect reporting regulations and guidance, assessing NRC staff needs for new construction-related training, and developing a framework to manage the impact of change on key new reactor oversight programs.

Challenge 4. Management of Regulatory Processes Associated with High-Level Radioactive Waste

Overview

NRC regulates high-level radioactive waste generated from commercial nuclear power reactors. High-level radioactive waste is either spent (used) reactor fuel when it is accepted for disposal or waste materials remaining after spent fuel is reprocessed. Because of its highly radioactive fission products, high-level radioactive waste must be handled and stored with care. Since the only way radioactive waste finally becomes harmless is through decay, which for high-level waste can take hundreds of thousands of years, the waste must be stored and finally disposed of in a way that provides adequate protection of the public.

The United States has entered a period where the national policy for storing, reprocessing, and disposal of spent nuclear fuel is being reexamined. With the prospect of spent nuclear fuel being stored at reactor sites for the foreseeable future due to the uncertainty surrounding a permanent repository for high-level radioactive waste, along with recent judicial action, NRC is being called upon to reevaluate its management approaches to the issues associated with long-term high-level radioactive waste storage.

In 2010, NRC updated its Waste Confidence Decision—affirming that spent nuclear fuel could be safely stored onsite at nuclear power plants until a permanent waste repository is built. However, on June 8, 2012, the U.S. Court of Appeals for the District of Columbia Circuit ruled that NRC's waste-confidence decision had not adequately addressed all environmental effects and thus violated the National Environmental Policy Act.

On August 13, 2013, the U.S. Court of Appeals for the District of Columbia Circuit issued a decision. The court ordered NRC to "promptly continue with the legally mandated licensing process" for the Department of Energy's application for authorization to construct a geologic repository for high-level nuclear waste at Yucca Mountain, Nevada, "unless and until Congress authoritatively says otherwise or there are no appropriated funds remaining."

An independent spent fuel storage installation (ISFSI) is an NRC-licensed facility designed and constructed for the interim storage of spent nuclear fuel and other radioactive materials associated with the spent fuel. An ISFSI typically consists of a concrete storage pad, storage containers (casks), and any support facilities. As of March 2013, there were 69 licensed ISFSIs in the United States.

Issues

- Ensure safe and secure interim storage for increasing quantities of high-level radioactive waste for the foreseeable future until a permanent repository for high-level radioactive waste is operational.
- Address regulatory issues relative to a longer-than-anticipated time for interim storage of high-level waste.
- Maintain flexibility to address regulatory challenges related to the storage and transportation of spent nuclear fuel and high-level waste.

Assessment

Because the U.S. Court of Appeals for the District of Columbia Circuit found that NRC violated the National Environmental Policy Act in issuing its

2010 update to the Waste Confidence Decision and Temporary Storage Rule, the Commission suspended all licensing activities that rely on the Waste Confidence Decision and Rule. NRC created a Waste Confidence Directorate within the Office of Nuclear Materials Safety and Safeguards to oversee the drafting of a new Waste Confidence Environmental Impact Statement and Rule. The Commission has instructed the Directorate to issue the final environmental impact statement and rule by no later than September 2014.

In FY 2011, OIG audited the safety aspect of NRC's oversight of ISFSIs and identified areas where the agency could improve. During FY 2013, NRC staff revised its inspection manual establishing a minimum frequency for conducting routine ISFSI safety inspections.

Challenge 5. Management of Information Technology

Overview

NRC needs to continue upgrading and modernizing its information technology (IT) capabilities to meet its IT/information management strategic goals. These goals include ensuring that NRC staff have quick and easy access to information, providing IT solutions that are easy to use and increase agency program performance, and delivering excellent service.

Issues

- Maintain and enhance information technology activities to strengthen the productivity, efficiency, and effectiveness of agency programs and operations.
- Expand mobile computing – "work from anywhere" – options such as remote access from NRC-issued laptops, non-NRC computers, and hand-held devices including a "bring your own devices" pilot program to allow for the distributed work locations of NRC staff.
- Improve information retrieval with better categorization and organization, enterprise content management, and improved search capabilities.
- Enhance "work with anyone" capabilities to include virtual meeting and collaboration tools with internal and external stakeholders including licensees and the public.

Assessment

In December 2010, the U.S. Chief Information Officer promulgated the 25-Point Implementation Plan to Reform Federal Information Technology Management. The plan tasked NRC and other agencies with undertaking specific management reforms and policy changes. During FY 2013, OIG audited NRC's compliance and recommended that NRC could improve IT management. As a result, the agency has agreed to institutionalize current training for its IT acquisition staff, create short-term and long-terms plans for its data centers, and include all data centers in future consolidation efforts.

Also during FY 2013, OIG evaluated NRC's progress in implementing social media at NRC. The evaluation found that NRC has made significant progress and is compliant with Federal social media policies and regulations. However, consistent with the fact that NRC is still in its early stages with its social media program, there remain areas where the agency can enhance its efficiency and effectiveness.

These areas include integrating social media into existing policies, training, and practices; implementing more social media specific security, training, and awareness safeguards; establishing a more prominent voice in the digital realm, and maximizing the potential of social media to enhance interaction with agency stakeholders and engage them in a dialogue on nuclear issues. The agency has implemented a number of recommendations such as soliciting input on a regular basis, making the authors more prominent in articles, and ensuring that information on how social media can be accessed by users is clearly defined and visible.

CHALLENGE 6. ADMINISTRATION OF ALL ASPECTS OF FINANCIAL MANAGEMENT AND PROCUREMENT

Overview

NRC faces significant challenges to efficiently, effectively, and economically manage its corporate resources within the parameters of its budget. In response to a challenging budget environment, NRC is pursuing strategies to target areas of inefficiency.

NRC must meet the requirements of several financial management statutes, including the Federal Managers' Financial Integrity Act and the Chief Financial Officers Act. These acts mandate NRC to establish controls that reasonably ensure that (1) obligations and costs comply with applicable law; (2) assets are safeguarded against waste, loss, unauthorized use, or misappropriation; and (3) revenues and expenditures are properly recorded and accounted for. These acts also encompass programmatic and administrative areas, as well as accounting and financial management.

NRC's procurement of goods and services is made with an aim to achieve the best value for the agency's dollars in a timely manner. Agency policy provides that these activities support the agency's mission; be planned, awarded, and administered efficiently and effectively; and be consistent with sound business practices and contracting principles. Agency efforts are currently focused on the goals of achieving

1) a 21st century acquisition program that uses state-of-the-art acquisition methodologies for acquisition planning, execution, management, and closeout, and
2) an acquisition program that fully integrates with the agencywide program and financial planning and budget execution.

Financial Management

- Respond to the current challenging budget environment.
- Improve the performance and functionality of the agency's core financial system.
- Improve the integration and functionality of all financially-related systems.

Procurement

- Implement a 21st century acquisition program that will interface with the agency's core financial management system.

Assessment

Financial Management

During FY 2013, NRC continued to demonstrate sound financial management practices. The *Audit of the NRC's Financial Statements for Fiscal Year 2012* resulted in an unqualified audit opinion. Moreover, other

OIG audits demonstrated that NRC is in compliance with Federal laws and standards related to financial management.

The agency also continues its efforts to enhance controls over financial management and to reduce agency overhead costs. For example, NRC established a Transforming Assets into Business Solutions Task Force to analyze and assess NRC's business practices and develop a plan to reduce the duplication of efforts in corporate and office support areas. However, OIG audits identified additional opportunities for improvement in financial operations.

An FY 2013 audit on NRC's process for calculating license fees indicated that NRC has opportunities to improve management of fees. Specifically, the audit recommended that the agency develop an annual validation process to compare budget estimates to actual costs and make adjustments as needed to the fee calculation process.

During an FY 2013 audit of travel charge cards, OIG found that NRC has an opportunity to maximize NRC's rebates by using recommended Federal strategies. The audit recommended that the agency implement a policy to pay centrally billed travel card accounts daily or weekly and explore the possibility of using another agency's existing task order to obtain a higher rebate rate.

During FY 2013, OIG Investigations continued to see examples where NRC employees misused their Government travel charge cards, for example, by charging items not associated with official travel. While NRC has made efforts to address this area, continued efforts are needed to reduce and eradicate misuse.

Procurement

NRC continues to upgrade its procurement system to streamline the agency's contracting practices. During FY 2013, OIG monitored the agency's procurement activities through meetings with the Division of Contracts, Office of Administration.

During FY 2014, these meetings will focus on the agency's Strategic Acquisition System scheduled to be implemented in the August – October 2013 time period.

During FY 2013, an OIG investigation concluded that a contractor improperly billed the agency and failed to provide adequate deliverables. A settlement was reached where the contractor agreed to pay damages plus investigative costs in the amount of $456,938. While NRC program office and contracting officials rejected deliverables and denied charges from the contractor, this OIG investigation highlights the need for continual vigilance in contract administration.

CHALLENGE 7. MANAGEMENT OF HUMAN CAPITAL

Overview

For several years, NRC experienced significant workforce growth resulting from increased interest in nuclear power. During FY 2013, NRC‘s workforce was approximately 4,000 staff positions. Going forward, NRC will need to support increasing mandates in a challenging budget environment. NRC must institutionalize an approach that focuses on its mission of protecting the public health and safety while remaining mindful of staff needs. To manage human capital effectively, while continuing to accomplish the agency's mission, NRC must continue to implement initiatives in the following areas:

- Reduce inefficiencies and overhead by centralizing and streamlining processes while maintaining or improving the level of service to agency offices.
- Space planning.

Issues

- Respond to a challenging budget environment.
- Adapt training and development programs to the changing needs of agency staff.
- Facilitate continuation of space consolidation efforts.
- Address knowledge management in light of the high number of senior experts and managers who are or will be eligible to retire.
- Enhance the environment for raising concerns.

Assessment

NRC, like many other Federal agencies, is dealing with the need to become more efficient in performing its mission. In response, NRC has developed a comprehensive human capital management system that is consistent with the agency's core values. The Office of the Chief Human Capital Officer (OCHCO) has implemented specific hiring controls in order to address the agency's projected shortfall in salaries and benefits. Additionally, OCHCO developed long- and short-term staffing plans focusing on mission-related work. These plans are effective tools for offices over the next several years as they balance grade structure, manage positions and achieve salary and benefit targets.

During audits and evaluations, OIG considers both budget information for NRC programs and training needs for staff and makes recommendations, as appropriate, for improvements in these areas. For example, in FY 2013, OIG conducted audits that discussed improvements needed in the training offered by the agency concerning NRC's (1) budget execution process and (2) travel charge card program.

NRC and the General Services Administration continue to work together to obtain approval for a consolidated housing plan for NRC headquarters that meets current standards for space utilization. After approval by the Office of Management and Budget, the plan will be presented to Congress. The final decision will rest with the House Committee on Transportation and Infrastructure and the Senate Committee on Environment and Public Works.

NRC continues to make progress on reconsolidation of its staff from headquarters interim buildings to the White Flint Campus' Two White Flint North and Three White Flint North buildings.

During FY 2013, OIG also reported on the agencywide safety culture and climate on the basis of an employee survey taken in September 2012.[4] When compared to previous surveys, agency staff rated 8 of 19 categories less favorable than in 2009.

Some of these areas include the staff's comfort level in elevating concerns and raising differing opinions, staff development and management, agency image, and the impact of metrics on work quality. Regarding knowledge management, the survey identified that NRC has made improvements in this area but still has opportunities to improve. The survey also indicated that staff may be less comfortable raising concerns since the previous survey was conducted. NRC management has assembled a group consisting of agency staff from multiple offices to seek opportunities for improvement in these areas.

In FY 2013, OIG investigative activities highlighted various human capital issues. For example, OIG completed an investigation regarding concerns associated with NRC's "open door" policy and Differing Professional Opinion (DPO) program. Several interviewees told OIG that because there is a perception by NRC staff that filing a DPO leads to retaliation, many staff are unwilling to use the program. Additionally, OIG investigations completed work and reported instances of misuse of the Transit Subsidy Benefits Program and undocumented Government overtime and time and attendance.

CONCLUSION

The seven challenges contained in this report are distinct, yet are interdependent to accomplishing NRC's mission. For example, the challenge of managing human capital affects all other management and performance challenges.

The agency's continued progress in taking actions to address the challenges presented should facilitate achievement of the agency's mission and goals.

Scope and Methodology

This report presents the IG's annual assessment of the most serious management and performance challenges facing the NRC. The challenges represent critical areas or difficult tasks that warrant high level management attention.

To accomplish this work, the OIG focused on determining (1) current challenges, (2) the agency's efforts to address the challenges during FY 2013, and (3) planned agency efforts to address the challenges.

OIG reviewed and analyzed pertinent laws and authoritative guidance, agency documents, and OIG reports, and sought input from NRC officials concerning agency accomplishments relative to the challenge areas and suggestions they had for updating the challenges. Specifically, because challenges affect mission critical areas or programs that have the potential to impact agency operations or strategic goals, NRC Commission members, offices that report to the Commission, the Executive Director for Operations, and the Chief Financial Officer were afforded the opportunity to share any information and insights on this subject.

OIG staff conducted this assessment from May through August 2013 at NRC headquarters.

End Notes

[1] Special nuclear material, as defined by Title 1 (Section 11. Definitions) of the Atomic Energy Act of 1954, is plutonium, uranium-233, or uranium enriched in the isotopes uranium-233 or uranium-235.

[2] FISMA is a United States Federal law enacted in 2002. The act requires each Federal agency to develop, document, and implement an agencywide program to provide information security for the information and information systems that support the operations and assets of the agency.

[3] An authorization to operate is a formal designation that authorizes operation of a business product and explicitly accepts the risk to agency operations.

[4] 2012 NRC Safety Culture and Climate Survey, OIG-13-A-15, March 29, 2013.

[5] Inspector General's Assessment of the Most Serious Management and Performance Challenges Facing NRC.

INDEX

G

H

I

J

Q

R

S

T

U

V

W